T0192338

Power Analysis of Trials with Multilevel Data

CHAPMAN & HALL/CRC

Interdisciplinary Statistics Series

Series editors: N. Keiding, B.J.T. Morgan, C.K. Wikle, P. van der Heijden

Published titles

AGE-PERIOD-COHORT ANALYSIS: NEW MODELS, METHODS, AND EMPIRICAL APPLICATIONS Y. Yang and K. C. Land

ANALYSIS OF CAPTURE-RECAPTURE DATA R. S. McCrea and B. J. T. Morgan

AN INVARIANT APPROACH TO STATISTICAL ANALYSIS OF SHAPES S. Lele and J. Richtsmeier

ASTROSTATISTICS G. Babu and E. Feigelson

BAYESIAN ANALYSIS FOR POPULATION ECOLOGY R. King, B. J. T. Morgan, O. Gimenez, and S. P. Brooks

BAYESIAN DISEASE MAPPING: HIERARCHICAL MODELING IN SPATIAL EPIDEMIOLOGY, SECOND EDITION A. B. Lawson

BIOEQUIVALENCE AND STATISTICS IN CLINICAL PHARMACOLOGY S. Patterson and B. Jones

CLINICAL TRIALS IN ONCOLOGY, THIRD EDITION S. Green, J. Benedetti, A. Smith, and J. Crowley

CLUSTER RANDOMISED TRIALS R.J. Hayes and L.H. Moulton

CORRESPONDENCE ANALYSIS IN PRACTICE, SECOND EDITION M. Greenacre

DESIGN AND ANALYSIS OF QUALITY OF LIFE STUDIES IN CLINICAL TRIALS, SECOND EDITION D.L. Fairclough

DYNAMICAL SEARCH L. Pronzato, H. Wynn, and A. Zhigljavsky

FLEXIBLE IMPUTATION OF MISSING DATA S. van Buuren

GENERALIZED LATENT VARIABLE MODELING: MULTILEVEL, LONGITUDINAL, AND STRUCTURAL EQUATION MODELS A. Skrondal and S. Rabe-Hesketh

GRAPHICAL ANALYSIS OF MULTI-RESPONSE DATA K. Basford and J. Tukey

INTRODUCTION TO COMPUTATIONAL BIOLOGY: MAPS, SEQUENCES, AND GENOMES M. Waterman

MARKOV CHAIN MONTE CARLO IN PRACTICE W. Gilks, S. Richardson, and D. Spiegelhalter

MEASUREMENT ERROR ANDMISCLASSIFICATION IN STATISTICS AND EPIDEMIOLOGY: IMPACTS AND BAYESIAN ADJUSTMENTS P. Gustafson

MEASUREMENT ERROR: MODELS, METHODS, AND APPLICATIONS J. P. Buonaccorsi

MEASUREMENT ERROR: MODELS, METHODS, AND APPLICATIONS J. P. Buonaccorsi

Published titles

MENDELIAN RANDOMIZATION: METHODS FOR USING GENETIC VARIANTS IN CAUSAL ESTIMATION S.Burgess and S.G.Thompson

META-ANALYSIS OF BINARY DATA USINGPROFILE LIKELIHOOD D. Böhning, R. Kuhnert, and S. Rattanasiri

POWER ANALYSIS OF TRIALS WITH MULTILEVEL DATA M. Moerbeek and S. Teerenstra

STATISTICAL ANALYSIS OF GENE EXPRESSION MICROARRAY DATA T. Speed

STATISTICAL AND COMPUTATIONAL PHARMACOGENOMICS R. Wu and M. Lin

STATISTICS IN MUSICOLOGY J. Beran

STATISTICS OF MEDICAL IMAGING T. Lei

STATISTICAL CONCEPTS AND APPLICATIONS IN CLINICAL MEDICINE J. Aitchison, J.W. Kay, and I.J. Lauder

STATISTICAL AND PROBABILISTIC METHODS IN ACTUARIAL SCIENCE P.J. Boland

STATISTICAL DETECTION AND SURVEILLANCE OF GEOGRAPHIC CLUSTERS P. Rogerson and I. Yamada

STATISTICS FOR ENVIRONMENTAL BIOLOGY AND TOXICOLOGY A. Bailer and W. Piegorsch

STATISTICS FOR FISSION TRACK ANALYSIS R.F. Galbraith

VISUALIZING DATA PATTERNS WITH MICROMAPS D.B. Carr and L.W. Pickle

Chapman & Hall/CRC
Interdisciplinary Statistics Series

Power Analysis of Trials with Multilevel Data

Mirjam Moerbeek

Utrecht University, The Netherlands

Steven Teerenstra

Radboud University Medical Center, The Netherlands

CRC Press is an imprint of the
Taylor & Francis Group, an **informa** business
A CHAPMAN & HALL BOOK

CRC Press
Taylor & Francis Group
6000 Broken Sound Parkway NW, Suite 300
Boca Raton, FL 33487-2742

First issued in paperback 2020

ISBN-13: 978-1-4987-2989-5 (hbk)
ISBN-13: 978-0-367-78344-0 (pbk)

Library of Congress Cataloging-in-Publication Data

Moerbeek, Mirjam, 1973-
Power analysis of trials with multilevel data / Mirjam Moerbeek and Steven Teerenstra.
pages cm. -- (Chapman & Hall/CRC interdisciplinary statistics ; 35)
"A CRC title."
Includes bibliographical references and index.
ISBN 978-1-4987-2989-5 (alk. paper)
1. Mathematical statistics. 2. Statistics--Methodology. I. Teerenstra, Steven, 1972- II. Title.

QA276.M5957 2015
519.5--dc23 2015010075

Visit the Taylor & Francis Web site at
http://www.taylorandfrancis.com

and the CRC Press Web site at
http://www.crcpress.com

Contents

List of figures

List of tables

Preface

What is the most effective method to lose body weight: a diet, a diet combined with physical exercise or a diet combined with a peer pressure group? Do special training programs increase unemployed people's chances of finding jobs on the labor market? Does a hormone treatment have an effect on the psychosocial functioning of children with growth retardation? Such questions are asked daily, not only by scientists but also by the general public. The similarity of these questions is that they focus on effective methods to delay or prevent disease or some unwanted and undesirable behavior, opinion or condition.

To compare various alternative methods in an experimental setting, it is important to seek a design that results in the highest statistical power at the lowest costs. It is common knowledge that the power level of a statistical test increases with increasing sample size. A sample size that is too small may result in not being able to detect an effect of a treatment, while an excessive sample size may be a waste of time and money of both the researchers and the trial's participants. It is therefore important to carefully calculate the required sample size before a trial is conducted.

Many textbooks and computer programs are of aid to the researcher who is planning a trial. These books and programs mainly apply to trials where the outcome measurements on disease, behavior or condition of a given subject are uncorrelated to those of other subjects in the trial. This assumption is very likely to be violated in trials where subjects are nested within groups, such as in multicenter clinical trials with patients nested within clinics and school-based smoking prevention interventions with pupils nested within schools. In the latter example, a pupil's smoking behavior will be influenced not only by the treatment condition he or she is assigned to, but also by the smoking behavior of other pupils in the same school, the teachers' smoking behavior and the school's policy toward smoking.

Ignoring the nested data structure may result in a study that is either under- or overpowered. Moreover, while the sample size affects a trial's power level, the allocation of units is also of importance. In the school-based smoking intervention, for instance, one has to decide whether to sample many schools and include just a few pupils per school, or to sample few schools and include many pupils per school.

In the past decades many journal papers that focus on statistical power analysis and optimal design for trials with nested data structures have appeared. It is now time to compile the published findings in these papers in one single book, and to present a related computer program to perform the power calculations. This is the book that you currently hold in your hands.

The two authors have traveled a long way before we were able to write this book. During this journey various people have been of invaluable support to our careers in the field of applied statistics.

Mirjam Moerbeek. My journey started in spring 1996, when I was appointed PhD researcher at the Department of Methodology and Statistics at Maastricht University, the Netherlands. Under the supervision of Martijn Berger and Gerard van Breukelen I wrote my PhD thesis on the design and analysis of multilevel intervention studies. Part of this research was done in collaboration with Weng Kee Wong and conducted at the Department of Biostatistics at the University of California at Los Angeles. During my PhD research, I frequented meetings of the Netherlands' Multilevel Modeling group, which were organized by Tom Snijders and Cora Maas. Both of them may be considered pioneers of multilevel modeling in the Netherlands.

After receiving my PhD, I continued my research on optimal design and power analysis at the Department of Methodology and Statistics at Utrecht University, the Netherlands. Peter van der Heijden encouraged me to apply for a prestigious research grant from the Netherlands Organization of Scientific Research (NWO). I was awarded a Veni grant in 2003 to extend my research on optimal design for trials with multilevel and longitudinal data, and a Vidi and Aspasia grant in 2008 to study optimal designs for discrete-time survival analysis and to write this book.

The Department of Methodology and Statistics at Utrecht University is the best place to conduct research like this because it has an enthusiastic group of researchers in the field of multilevel analysis, among whom are Joop Hox, Rens van de Schoot and Leoniek Wijngaards-de Meij. Together we organize the biennial International Conference on Multilevel Analysis.

Part of the research in this book was done by former PhD students of mine: Elly Korendijk studied the effects of a misspecification of the intraclass correlation coefficient in the design phase of a cluster randomized trial and Esther Oomen-de Hoop studied the design and analysis of stepped-wedge cluster randomized trials. Charlotte Rietbergen, a former master student of mine, studied crossover in cluster randomized trials. Sander van Schie, also a former master student, studied sample size re-estimation in cluster randomized trials and prepared an overview of papers with estimates of the intraclass correlation coefficient. Katarzyna Jóźwiak wrote the SPA-ML computer program to perform many of the sample size and power calculations that are presented in this book.

Steven Teerenstra. What I like about statisticians is their diversity of origins. My interest in statistics arose when I was looking for a more practically oriented job after finishing my appointment as a PhD student in mathematics in Nijmegen. At that time (2002) I had a freelance job in the Radboud

University Nijmegen Medical Center to build a research database and I was asked to also think about the statistical analysis of the data. I contacted a former colleague of mine, Wiebe Pestman, who had already travelled the road from mathematics to statistics. He gave me not only statistical advice but also suggested I become a statistician. Without formal statistical training or working experience, my first application outside Nijmegen was not successful. However, opportunities were closer by than I thought. After attending a talk at the department of Epidemiology, Biostatistics and HTA in Nijmegen (now the Department for Health Evidence), I had a chat with Gerhard Zielhuis, who told me that the biostatistics group was looking for a candidate. Under guidance of George Borm, my journey in biostatistics began: hands-on and steep. Because much of my consultancy involved cluster randomized trials, George encouraged me to do research in this topic. A choice that I enjoy till today, because the techniques also gave me insight in other fields of statistics such as longitudinal data and meta-analysis.

While becoming acquainted with the literature in the field, I came across Mirjam's papers. I saw that we had common interests and contacted her to collaborate, which led to our joint endeavors. A research grant by NWO allowed me to work with Anouk Spijker to investigate the efficiency of ANCOVA analysis of cluster randomized trials. With a grant from the Radboud University Nijmegen Medical Center, I engaged Esther Oomen-de Hoop as a PhD student to investigate practical solutions for cluster randomized trials with few clusters.

Mirjam Moerbeek, Utrecht, the Netherlands
Steven Teerenstra, Nijmegen, the Netherlands

1

Introduction

The scientific community performs tens of thousands of experiments each year to improve the treatment of diseases and psychological and social conditions with the aim to increase quality of life and life expectancy. Examples include substance use prevention and cessation interventions, trials that compare cognitive behavioral group therapy to pharmacotherapy for the treatment of social phobia, and trials that evaluate the effects of growth hormones on the psychosocial functioning of children with growth retardation.

The process of data collection can take up to several decades in the case of long-term interventions. In addition, efforts should be paid to good data analysis, including the use of an appropriate statistical model and proper treatment of missing data arising from drop-out and non-response. It is therefore of utmost importance to plan an experiment in such a way that the best treatments are selected with highest probability.

In the design phase of a trial, the treatments to be compared are selected, and choices are made on eligibility criteria, the duration of the study and the best training of professionals who are delivering the treatments to their patients or clients. Another important choice in the design phase is the required number of subjects, as the probability to detect a significant difference between treatment conditions depends on the chosen sample size. This probability is called the statistical power and the calculation of the required sample size to achieve a desired power level is called a power analysis.

In many trials in the social and biomedical sciences, the data have a so-called hierarchical structure, meaning that subjects are nested within groups. For such trials, not only the total number of subjects needs to be determined, but also their allocation over the groups. Should few groups with many subjects each be sampled, or many groups with few subjects each?

Similar questions may be asked for the design of longitudinal trials with repeated measurements over time that are nested within subjects. Should few subjects be sampled and should they be measured often, or should many subjects be sampled and they be measured just a few times?

Guidelines for sample sizes for trials with hierarchical data structures have been studied extensively in the last two decades. The aim of this book is to provide formulae to perform power calculations to social and biomedical researchers and to statisticians working in these fields of science. This first chapter serves as an introduction to basic concepts with respect to experimentation, hierarchical data structures, and study design and further elucidates the need for power calculations. A further description of the aim and contents of the book is given at the end of this chapter.

1.1 Experimentation

This section aims to provide an overview of the basic concepts of experimentation: controlling, randomization, stratification or blocking and matching, replication and blinding. Although these concepts appear obvious nowadays, their formal introduction was initiated less than a hundred years ago by the work of Fisher in the field of agriculture in the 1920s and 1930s (Fisher, 1926, 1935). The first modern randomized controlled trial in health care research is generally considered to be that of the Medical Research Council (1948), that compared the effect of streptomycin and bed rest to that of bed rest only on the treatment of pulmonary tuberculosis. The first randomized controlled trial in social research appears to have been conducted in the late 1920s (Forsetlund, Chalmers, & Bjørndal, 2007). For a more thorough introduction to experimentation the reader is referred to Jadad and Enkin (2007) and Torgerson and Torgerson (2008).

To evaluate the effect of a new type of therapy, surgery, teaching method or any other type of treatment, the measurements on outcome variables of subjects in the new treatment group have to be compared to those receiving a standard treatment, no treatment at all, or those who are assigned to a waiting list. The latter groups of subjects are often called control groups, and the corresponding treatment condition is called the control condition. Including a control is important since it enables the researcher to compare the performance of the new treatment to the performance of a standard treatment or no treatment at all. Without the control group, a researcher is unable to tell whether the performance of the new treatment is better than, worse than or equal to that of the standard or no treatment. In other words, the control serves as a benchmark and enables the estimation of the treatment effect, which is the difference in performance of the two treatment conditions on the outcome variable.

If one wishes to generalize the findings to all subjects within the population, a random draw should be made from this population. To achieve a fair comparison of treatment conditions, the treatment groups should be as similar as possible at baseline with respect to all measured and unmeasured covariates, which are variables that are supposed to have an effect on the outcome. For instance, as parents' smoking behavior has an effect on adolescent smoking behavior, it is important that the baseline percentages of adolescents whose parents smoke are equal over the treatment conditions in a smoking prevention intervention that targets adolescents.

For large sample sizes, this can be achieved by randomly assigning adolescents to treatment conditions and the design is called a completely randomized design. For small sample sizes, random assignment does not guarantee balance with respect to parent smoking behavior. This can be resolved by stratifying adolescents with respect to their parents' smoking behavior: one stratum consisting of adolescents who have one or two parents who smoke, while adolescents in the other stratum do not have a parent who smokes. Randomization to treatment conditions is then done within strata and the design is called a stratified randomized design. Alternative terminology is to use blocking, blocks and randomized block design wording instead of stratifying, strata and stratified randomized design. The difference is that the terminology of blocking, blocks and randomized block design is more often used for factors that are generally determinable and often controllable in the experiment and for which the sample size can be determined upfront, and called blocking factors (e.g., temperature).

The terms stratifying, strata and stratified randomized design apply more often to factors that cannot be determined upfront (e.g., concomitant medication use). Often the distinction is not clear-cut. An option that is more useful when multiple covariates are present is the matched-pair design. With this design, pairs of subjects who are as similar as possible with respect to all covariates are formed, and randomization is done within each pair. Special algorithms are available to balance treatment groups with respect to multiple covariates; see for instance G. F. Borm, Hoogendoorn, Den Heijer, and Zielhuis (2005). If such balancing methods have not performed well, or if these methods were not used at all, then the method of statistical control can be used to correct for between-group differences. This is achieved by including important covariates as predictor variables in a regression or analysis of covariance model.

Whichever strategy is chosen to prevent imbalance or to correct for it, it is important that relevant covariates are defined in the planning of a trial to be able to measure them in the data collection phase. It is therefore necessary to search the literature for relevant covariates prior to data collection.

Several replications should be made to estimate the variability of the error terms in a statistical model. That is, more than one subject should be available within each combination of treatment condition and covariate values. In a trial that randomizes to treatment conditions within gender, for instance, at least two males and two females are required in the control condition, and another two males and two females in the experimental condition. Replication should not be confused with taking repeat measurements on the same subjects. If repeated measurements over time are taken on each subject, then time should be included as a predictor in the statistical model.

In an ideal case, the trial is double-blind, meaning that neither the researchers nor the subjects know to which treatment condition each subject is assigned. Only a third party has access to the key that identifies to which condition each subject is randomized. With single blinding only the study participants are unaware to which treatment condition they are assigned. Double blinding is a means to lessen observer bias that occurs when the results are influenced by conscious or unconscious bias on the part of the observer. Such bias occurs when observers tend to adjust their scores on a subject's outcome variables to their expectations or judgments of the (known) treatment condition to which a subject is assigned.

Another reason for double blinding is to reduce selection bias which may occur when recruiters know beforehand the treatment to which a subject is assigned. The risk of both biases is illustrated by the finding that the outcomes and characteristics of included participants may be different in blinded trials than in unblinded trials (Veerus, Fischer, Hakama, & Hemminki, 2012) and hence blinding may influence both internal and external validity.

Unfortunately, blinding is not always an option. Double blinding is often used in pharmaceutical trials where different substances are compared. The pills or injections containing these substances should be as similar with respect to size, color, smell, weight and so forth as possible. In social and behavioral trials, treatment often relies on interpersonal interactions, such as risk-reduction sessions, peer pressure groups and training programs and blinding is often not possible.

The degree to which a researcher has control over the environment in which the experiment is conducted determines whether the experiment is a laboratory or field experiment. In a laboratory (pure) experiment, the control over the environment is maximal and all subjects are exposed to the same influences expect for the treatment condition to which they are assigned. Consider as an example a trial that investigates the effect of the opponent's ethnicity on the aggressiveness of players of dual-player computer games. Participants are assigned to an opponent of the same or different ethnicity and all other sources of variation are kept under the experimenter's control. Thus, the participants play the same computer game in rooms with the same background music, temperature, lighting, accessibility to food and drinks and so forth.

Laboratory experiments are often artificial and doubts on generalization of results to the real world exist. In a field experiment, the daily life environment is treated as the laboratory. Randomization to treatments is under experimental control but the experimenter does not have an influence on uncontrolled events. This is not considered a problem as long as both treatment groups are equally exposed to uncontrolled events. However, a problem arises when the one treatment group is exposed to uncontrolled events to a higher degree than the other group and/or when the uncontrolled events interact with treatment condition. In both situations, the external factors influence the one group more than the other, thereby threatening the validity of the study. Consider as an example the effects of exercise on the reduction of body weight in a controlled trial. A difference in body weight between both treatment groups can only be attributed to the effects of exercise if the calorie intake is equal across groups.

1.1.1 Problems with random assignment

Although randomization is a strong tool to achieve treatment groups that are equal with respect to relevant covariates, it also has some drawbacks. In the ideal case, randomization is done by the researcher or an independent statistician, but in practice it is often done by those who control access to potential participants, such as general practitioners, therapists and school principals. They often tend to assign the participants they consider most deserving or most likely to benefit to the new treatment group, a process which is called selection bias.

People who are recruited to participate in an experiment are often only willing to do so if they are assigned to the interesting and promising new treatment. Some of them will even try to force the person performing the randomization to assign them to the new treatment, resulting in a treatment group that consists of more demanding or more assertive subjects than the control condition, a situation that is most undesirable when the outcome variable is related to these qualities of character.

When participants cannot influence the randomization procedure, they can still undermine the study. For instance, those assigned to the control condition can try to benefit as much from the new treatment as possible by contacting participants in the new treatment group. This is especially a threat if the new treatment condition consists of oral or written information to improve lifestyles, and less so when the new condition consists of medication or surgery. This process is called control group contamination since information on the new treatment somehow leaks to participants in the control group. In addition, drop-out rates among those in the control group are often higher than those in the new treatment group and disappointment may be common after allocation to the control condition (Lindström, Sundberg-Petersson, Adami, & Tönnesen, 2010).

Resentment of those in the control condition can be lessened by using a placebo, which is an irrelevant treatment in the control group, such as a sugar pill in pharmaceutical trials. As an alternative, one can inform the participants in the control group on the necessity of random assignment and the use of a control group and assure them of having one of the more desirable treatments at the end of the trial, provided these treatments have a relevant effect and do not show harmful side effects.

A related problem occurs when participants in the new treatment condition do not wish to participate if they find this new treatment undesirable, if they expect its effects are negligible or if they expect unwanted or harmful side effects. For instance, parents who smoke might not be interested in their children participating in a smoking prevention intervention. This problem may be solved to only include those participants who have expressed their willingness to participate. As is obvious, this may induce problems with respect of generalization to the general population.

In some cases randomization is not possible due to ethical aspects. For instance, a general practitioner might not be willing to let only half of his patients benefit from the new and promising treatment. In other cases, one cannot force subjects to adhere to the treatment assigned. For instance, one cannot force subjects to smoke or not in a trial on the effects of smoking on lung cancer. In such cases, treatments can only be compared on basis of a quasi-experimental design. With this design, the experimenter cannot randomly assign subjects to treatment conditions and must rely on existing differences between people with respect to the variable of interest. The pitfall is that this variable may be correlated with covariates that also have an effect on the outcome and an exhaustive literature review to detect all such covariates is therefore required.

The sample size formulae that are presented in the next chapters are valid for experimental designs where randomization is done in such a way that the treatment groups are comparable with respect to covariates. In Section 5.2 the effects of matching and pre-stratification on power are treated in further detail.

1.2 Hierarchical data structures

In many experiments in the social and biomedical sciences, the data have a so-called hierarchical structure.This means that subjects are nested within groups that may themselves be nested within higher order groups, and so on. The words *nested*, *clustered* and *multilevel* are often used as synonyms to the word *hierarchical*. Examples are clinical trials with patients nested within clinics, school-based smoking and substance abuse prevention interventions with pupils nested within classes within schools, and studies in the field of clinical psychology with clients nested within therapists.

Hierarchical data structures arise from multistage sampling where sampling is done in a number of successive steps. For instance, therapists are sampled from a population of therapists in the first step. In the second step, a sample of clients is taken from therapists selected in the first step. Only those patients whose therapists were selected in the first step have a chance of enrolling into the trial; the changes of the other patients decline to zero once the first step is executed. Thus, selection probabilities are not constant when multistage sampling is used. The sampling scheme for two-stage sampling is illustrated in Figure 1.1. The large circles represent the therapists; the small circles represent their clients. The black large circles are therapists who are not selected in the first step, and neither are any of their patients in the second step, as represented by the small black circles. The white large circles represent therapists selected in the first step. For each of these, some of his or her patients are selected in the second step and some of them are not, as represented by the small white and black circles, respectively.

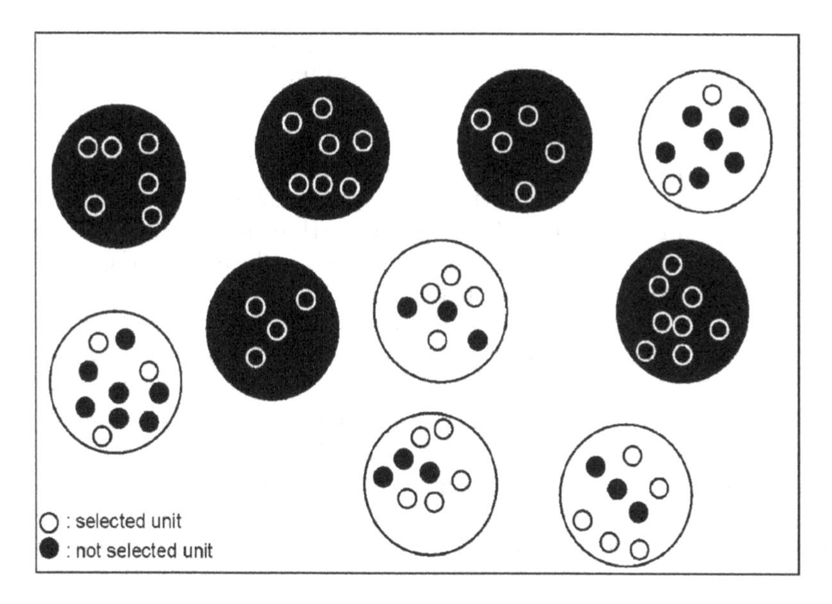

Figure 1.1: Graphical representation of a two-stage sampling scheme.

The advantage of the two-stage sampling scheme is that it is often less expensive than taking a simple random sample where each patient has equal changes of being selected. This is explained by the fact that the number of therapists is often lower if multistage sampling is used, meaning that fewer therapists have to be communicated with and fewer therapists have to be trained to deliver the new type of therapy to their clients. In addition, the patients of the same therapists are geographically organized, which may reduce travel costs.

Similar arguments hold for longitudinal studies where repeated measurements are nested within subjects. With such studies it may be more convenient to follow the same subjects over time than to take a new sample of subjects at each time point. In other words, a cohort design may be more convenient than a cross-sectional design.

The other side of the coin is that data analysis and power calculations for trials with hierarchical data are generally more complicated than those for trials without nesting because measurements of subjects within the same group are likely to be correlated. In other words, a subject's measurements on health, behavior, attitude, opinion and other relevant outcomes cannot be regarded as independent of those from other subjects within the same group. Such dependent data may arise from therapist effects, which occur when the success of treatment depends on the skill and training of the person delivering the treatment, or on the quality of the equipment that is used by this person. This is often the case in trials that rely on surgery, interviewing or physical or psychological treatment where subjects treated by more experienced or more empathic health professionals are more likely to recover.

Correlated data structures may also arise from mutual influence between subjects within the same group. In school-based smoking prevention interventions, for example, a pupil's smoking behavior is likely to be influenced by behaviors of other pupils within the same class. In addition, the group environment may also be a cause of dependency within the same group. In the latter example, a pupil's smoking behavior may be influenced by the smoking behavior of teachers, the school policy toward smoking and the availability of cigarettes in the school and its neighborhood.

Genetic factors may cause the occurrence of certain diseases and conditions within families. For instance, some families are more prone to obesities and breast cancers than others, resulting in correlated outcomes on disease occurrence within the same family. Finally, dependency may occur when people select the groups they belong to. People select their friends on basis of mutual interests or shared opinions and attitudes, while parents select their children's schools on basis of shared appraisals of the school's system and values. Once the groups are established, they will become differentiated and both the group and its members will be influenced by group membership.

Whatever the cause for correlation in trials with hierarchical data structures, ignoring it may result in incorrect point estimates and standard errors of regression coefficients and hence incorrect conclusions with respect to the effect of treatment and relevant covariates. Several approaches for dealing with hierarchical data have been suggested in the past; see Moerbeek, Van Breukelen, and Berger (2003a) for an overview. One such approach is to represent the groups by dummy variables in the regression model that relates outcome to treatment condition and covariates in an analysis of covariance model. This model is a so-called fixed effects model, and the conclusions drawn on basis of such a model are only valid for those groups that are enrolled in the study.

If one wants to generalize the findings of the study to the whole population from which these groups are sampled, then one should use the multilevel model. This model is also known as the hierarchical model, mixed effects model, random coefficient model or variance component model. The multilevel model assumes the groups represent a random sample from the underlying population. Snijders and Bosker (2012) suggest using the multilevel model when the number of groups is at least twenty. With fewer groups, the information about the population provided by the groups is insufficient and cannot justify generalization of the results to the whole population.

The multilevel model differs from the traditional (single level) model since it explicitly accounts for the nested data structure by including random effects at the group level (and higher order levels, if present). These random effects represent the between group variation with respect to the outcome variable. The amount of variance in the outcome variable at the group level relative to the total variance is called the intraclass correlation coefficient. In some studies, the values of the random effects are of importance, for instance, in school-effectiveness studies, the random effects are used to rank schools with respect to their performance. In experiments, on the other hand, the random effects themselves are generally not of main importance and are therefore treated as nuisance parameters. Ignoring them, however, may result in incorrect point estimates and standard errors of model parameters.

These standard errors also play a crucial role in the calculation of the statistical power to detect a treatment effect, and hence in the calculation of the required sample size to achieve a desired power level. As will be shown in the following chapters, this standard error is a function of degree of correlation within a given group. This function, in turn, depends on the chosen design. Various designs with hierarchical data structures are available, as will be explained in the next section.

1.3 Research design

A research design is used to structure the research and data collection. It determines how the available financial and personnel resources should be used to answer the research question. The primary aim of an experiment is to estimate the effect of treatment as precisely as possible to achieve highest possible power for the test on treatment effect. Statistical power increases with the number of subjects. If the available sample size is insufficient and cannot easily be increased, then power may be enhanced by using a more efficient design, for instance a crossover design.

With such a design, subjects are given the treatment conditions in different orders over a series of periods such that the precision of the estimated treatment effect does not depend on the between-subject variation. So, the selection of the research design is a very important step in the planning phase of a trial. The optimal design provides highest possible power given the available resources and is still feasible from a practical point of view. The crossover design, for instance, cannot be used in the presence of carry-over effects that occur when the effect of treatment in a certain time period extends to subsequent time periods. A matched-pair design may be considered a more viable option in such cases as it is not hampered by carry-over effects.

Whichever design is chosen, it must be structured in such a way to allow the researcher to answer the research questions. In that sense, research design is a logical question rather than a logistical one (De Vaus, 2005). Research design does not indicate whether relevant outcome variables should be measured by means of open or structured interviews, questionnaires, observations or other method. Although the selection of the best method of data collection is an important one, it is outside the scope of this book.

The simplest design for an experiment is the simple randomized trial. With this design, subjects are randomly drawn from their population and assigned at random to the treatment or control condition. Many trials in the social and behavior sciences have a hierarchical data structure. This data structure may already be present prior to randomization to treatment conditions. In such cases, one may choose to randomize complete groups to treatment conditions or randomize subjects within groups. The formal terms for these two trials are cluster randomized and multisite trials, respectively. For instance, pupils are nested in schools prior to the implementation and the randomization of schools to treatment conditions does not determine to which school a pupil goes.

In other cases, the nesting of subjects within groups is a consequence of research design, and a subject's group membership may be established once randomization is done. This is the case in individually randomized group treatment trials where treatment is offered in groups. For instance, patients with chronic pain and high absenteeism from work may be randomized to learning groups. Such groups do not exist prior to the experiment and are established once randomization is done. These treatment groups will differentiate because of mutual influence of members within the same group and the skills and experience of the professional who leads the group.

A short overview of the main types of trials that are considered in this book is now given: cluster randomized trials, multisite trials, pseudo cluster randomized trials, individually randomized group treatment trials and longitudinal intervention trials. These are treated in more detail in Chapters 4 through 9. Chapter 10 shows power calculations for other trials with clustered data.

1.3.1 Cluster randomized trial

Cluster randomized trials are also referred to as group randomized trials, place-based trials or community intervention studies. With such trials, complete groups, such as schools, general practices and worksites are randomized to treatment conditions and all subjects within the same group receive the same treatment. This design is often chosen for financial and logistic considerations. Consider for instance a school-based smoking prevention intervention. It is less expensive to teach a class of pupils about the risks of smoking than to teach individual pupils who are not grouped within classes. In addition, administrative costs will be reduced since all pupils within the same school receive the same treatment. In some cases, cluster randomization is chosen over individual randomization from an ethical view, for instance, if it is considered unethical to treat some subjects and not others within the same group. The main problem with cluster randomized trials is that in general they need a larger sample size than simple randomized trials which will be fully explained in Chapter 4.

Figure 1.2 gives a graphical representation of a cluster randomized trial. The subjects represented in white are randomized to the control; those represented in black are randomized to the intervention. There are eight clusters (as

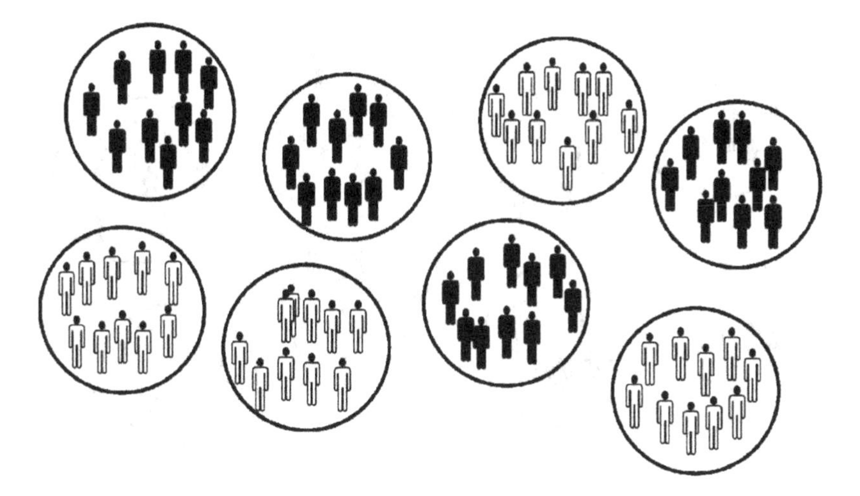

Figure 1.2: Graphical representation of a cluster randomized trial. (Reprinted with permission from Moerbeek, M., & Teerenstra, S. (2010)).

represented by circles) and each of them consists of ten subjects. There is only one treatment condition per cluster; hence the design is a cluster randomized trial.

1.3.2 Multisite trial

Multisite trials are also known as multicenter trials or multiclinic trials. Randomization of subjects to treatment conditions occurs within groups, so each treatment condition is available in each group. This design can be shown to have a higher power to detect a difference between treatments than the cluster randomized trial. In addition, it also allows for the estimation of group by treatment interaction.

The main drawback of multisite trials is that they may be subject to control group contamination, for instance as a result of interaction between subjects within the same group. In other trials, contamination may be due to the person delivering the intervention. When both the standard and new treatments or therapies are available from the same health professional it would be difficult for the professional not to let patients in the standard treatment group benefit from the new treatment. The multisite trial may be very suitable for double-blind pharmaceutical trials but less so for trials that rely on interpersonal relationships and communication. A graphical representation of a multisite trial is given in Figure 1.3. Both treatment conditions are equally represented within each cluster; hence the design is a multisite trial.

Figure 1.3: Graphical representation of a multisite trial. (Reprinted with permission from Moerbeek, M., & Teerenstra, S. (2010)).

1.3.3 Pseudo cluster randomized trial

In some trials the therapist or health professional (acting as the cluster) is requested to recruit eligible patients or clients. For such trials, cluster randomization may result in poor recruitment by those therapists or health professionals who are randomized to the less interesting control condition. In addition, cluster randomization may result in selection bias as the therapists and health professionals already know after recruiting one patient what the randomization outcome will be for all other patients they recruit and this knowledge may influence which patients they select. Randomizing patients of therapists or health care professionals to both conditions would alleviate these problems, but if this causes contamination, the multisite trial is not a good solution either.

For such trials, pseudo cluster randomization is proposed as an alternative to both cluster randomization and 50% − 50% randomization within clusters as in a multisite trial. With such trials both treatment conditions appear in each group in such a way that the amount of control group contamination is minimized. This is achieved by a randomization scheme as presented in Figure 1.4 and this will be further discussed in Chapter 7.

1.3.4 Individually randomized group treatment trial

A common characteristic of (pseudo) cluster randomized trials and multisite trials is that subjects are already nested within groups prior to randomization to treatment conditions. The randomization procedure does not determine the

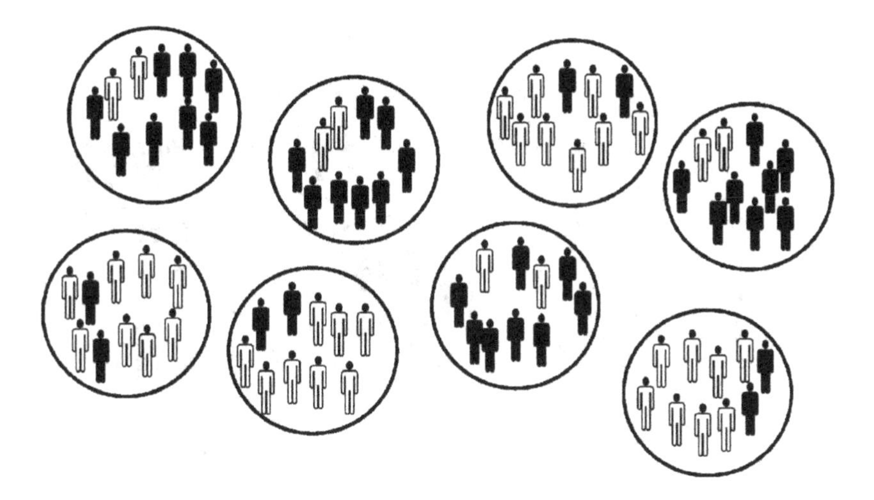

Figure 1.4: Graphical representation of a pseudo cluster randomized trial. (Reprinted with permission from Moerbeek, M., & Teerenstra, S. (2010)).

group to which each subject belongs, although subjects may change groups during the course of the study and thereby change treatment condition.

In individually randomized group treatment trials, on the other hand, the randomization procedure does determine each subject's group membership. Nesting may occur in both treatment conditions and for such trials it is not unlikely that the group size and intraclass correlation coefficients vary over treatment conditions. This artefact needs to be taken into account in the statistical model and power calculations. Nesting may also occur in just one treatment condition, and the control condition often consists of subjects who are assigned to a waiting list. The latter randomization scheme is presented in Figure 1.5.

1.3.5 Longitudinal intervention study

With longitudinal intervention studies, subjects are randomized to treatments and followed for some period of time. Measurements are taken at baseline and during the course of the study, for instance at the end of treatment and during follow-up. Depending on the type of intervention, measurements are taken some weeks, months or even years after the delivery of treatment. A subject's treatment condition does not change over time and the measurements within subjects are expected to be correlated. A major question for such studies is whether the change of the outcome variable over time varies over treatment conditions. Figure 1.6 is a graphical representation of a longitudinal intervention study based on four measurements within each subject. As before, white and black subjects represent individuals in the control and intervention groups

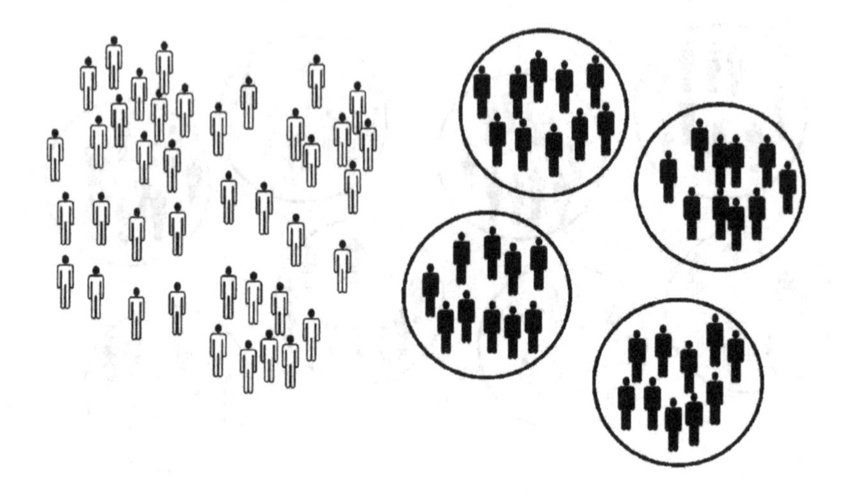

Figure 1.5: Graphical representation of an individually randomized group treatment trial. (Reprinted with permission from Moerbeek, M., & Teerenstra, S. (2010)).

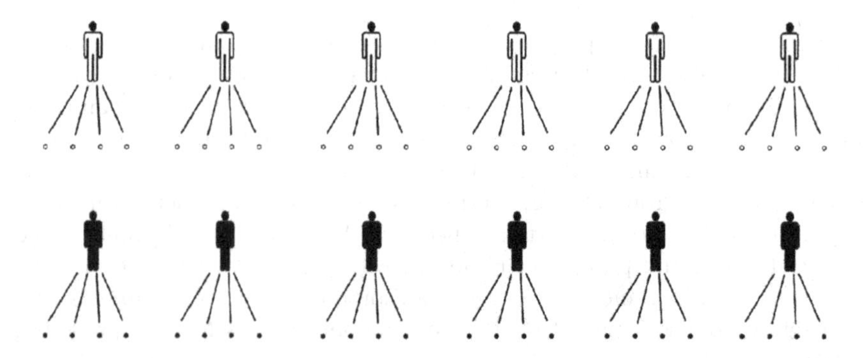

Figure 1.6: Graphical representation of a longitudinal intervention study.

and the same color scheme is used for the measurement occasions, represented by dots. It is clear subjects do not change their treatment condition during the course of the trial.

1.3.6 Some guidance to design choice

Based on what will be explained in more detail in the coming chapters, where the designs are discussed, some guidance to design choice can be obtained. In general it can be said that, where possible, within-unit comparisons are more efficient than between-unit comparison (i.e., crossover trials are more efficient than parallel group trials), all other things being equal. Also, where possible, randomizing units at lower levels is more efficient than randomizing at higher levels. For instance, a multisite trial (randomization at subject level) is more efficient than a cluster randomized trial (randomization at cluster level), all other things being equal.

In addition, including more units at the level where the randomization is increases power when no interactions are present, for instance, including more clusters in a cluster randomized trial or including more subjects per site in a multisite trial. Increasing the sample size at other levels will increase power as well, but often only to a certain point. For example, in a cluster randomized trial, increasing the number of clusters indefinitely will not always lead to a power of 80%. In case of interaction, it may be the goal to test the interaction and for this it is more useful to increase the sample size concerning the appropriate interaction factor. For example, in a multisite trial with site by treatment interaction, the power for the interaction test increases when more sites are included.

In many instances, the choice of design is limited due to the characteristics of the intervention, the population, the budget and/or time available and so on. For instance, if an intervention cannot be 'undone', a crossover trial is not possible. Also, if there is risk of contamination, a multisite trial is not the first choice. Furthermore, often only a few measurements are possible. In such cases where the choice is limited, the formulas discussed in the following chapters and the power analysis program allow us to compare the options.

Then there are situations where the designs discussed here do not fit well. In such cases it is wise and fun to see the design of your trial as a research question and devise with a statistical expert a design that fits your trial best. Chances are that you may invent a design that is worth adding to the array of design options.

1.4 Power analysis for experimental research

One reason for performing power analyses is that large-scale experiments in the social and biomedical sciences are often very expensive. First, there are the efforts of many scientific and administrative staff members. In addition, costs are associated with implementing treatments such as medication and surgery and incentives for professionals who deliver the treatments to patients, clients or laboratory animals.

In a scientific world with increasing competition for research funding, agencies that finance scientific research demand a thorough financial justification in research proposals. Part of this is a justification of the chosen sample size. Insufficient sample size is a waste of money since it may not be able to detect an effect of a new treatment, even though this effect exists in the population. Too large sample size is also a waste of money since it uses more subjects than necessary and thus results in a negligent use of the subjects' participation. In addition, it can raise ethical objections to expose more subjects to the intervention than necessary, for instance if the trial's subjects are laboratory animals that are killed for the purpose of the study. Therefore, accurate sample size calculations are indispensable in grant applications.

Another reason for performing sample size calculations is that it is good research practice. It forces researchers to think up front of the primary objective(s) of their research and especially decide the primary outcome variable(s) that reflect their primary objective(s). As the sample size does not only depend on the primary outcome(s) but also on the design chosen and the statistical analysis method, performing a sample size calculation makes researchers think a priori about their design and analysis models as well.

Finally, planning the sample size correctly also discourages unplanned looking at the data (while recruitment is still ongoing) to see whether there is already statistical significance. This is important, because such interim analyses could lead to a higher probability of chance findings. For all of these reasons, sample size planning enhances quality of research and has become part of reporting guidelines such as the CONSORT statement for parallel group randomized controlled trials (K. Schulz, Altman, & Moher, 2010; Moher et al., 2010) and the CONSORT statement for special kinds of trials, such as cluster randomized trials (M. K. Campbell, Elbourne, & Altman, 2004). Therefore, scientific journals demand sample size calculations to be included in methods sections of papers.

Choosing a sample size on basis of practical considerations such as a limited budget or a limited number of eligible subjects who can be recruited without motivation that this will probably lead to meaningful results, will not be accepted by funding agencies and will be questioned by scientific journals.

A common and reliable way to motivate sample size is a statistical power analysis. For a single primary objective expressible in one outcome variable, a power analysis calculates the sample size in such a way that a prespecified meaningful difference between treatment conditions can be detected with a predefined probability. This predefined probability is called the power (for detecting a difference of that size). A power analysis is therefore a way to ensure that a meaningful result of the study can be obtained with a high probability provided this meaningful difference is actually present.

In this sense, a power analysis avoids studies that are too small to detect any meaningful difference (underpowered studies) and studies that are so large that they can detect small differences that are not considered relevant (overpowered studies). A power analysis forces the researcher to decide on the minimal treatment effect that is considered relevant, the research design, the statistical analysis method, the maximum acceptable rate of false positives (type I error), and last but not least, the desired level of power. The power level of a trial should be chosen to reflect the consequences of not detecting a difference between treatment conditions.

For pharmaceutical companies, the consequences are a new drug not approved and the prospects of large profits canceled. Ethical committees tend to disapprove trials that have less than 80% power. A more lenient restriction on the minimal power level would result in smaller sized trials, which are often easier to initiate and complete. For instance, a trial with 50% power has approximately half the size of a trial with 80% power. The results of many such smaller studies can then be combined in a meta analysis. The common objection against this approach is that meta analyses are hampered by publication bias. Studies with non-significant results often remain unpublished and are excluded from meta analyses.

The results of meta analyses are therefore overly optimistic. As publication bias often applies to small studies, it is often advocated that each single trial has sufficient power of, say, 80%. G. F. Borm, Den Heijer, and Zielhuis (2009) have shown that this requirement may be relaxed. Even if each single trial has a power as low as 50% and the publication rate of nonsignificant results is 60%, the bias of the treatment as calculated from a meta analysis was at most 15%. In other words, the treatment effect as estimated in the meta analysis was at most 15% larger than the true treatment effect.

The fact that trials with 50% power may also contribute to science does not imply a power analysis may be omitted. Without such a power analysis, the power of a study may be low. So, whenever one wishes to draw conclusions based on one single large-sized trial or perform a smaller-sized trial that can subsequently be included in a meta analysis, an a priori power analysis must always be performed. The formal steps are detailed in Section 3.6.

Most power analyses are restricted to one primary objective expressible in one primary outcome variable. Usually, the power for other outcome variables is then much lower. For example, Tsang, Colley, and Lynd (2009) found in a review of randomized clinical trials powered on an efficacy outcome that power levels for adverse events could be lower than 10%. Power calculations for more primary objectives expressed in more than one primary outcome variable are also possible (G. Borm, van der Wilt, Kremer, & Zielhuis, 2007). For instance, if a study has two primary outcome variables, a safe way to obtain an overall power of the study at 80% is that each outcome variable should have at least 90% power (G. F. Borm, Houben, Welsing, & Zielhuis, 2006).

Although the need for sufficient power is commonly acknowledged, published studies may still be underpowered. An indication for that is that Charles, Giraudeau, Dechartres, Baron, and Ravaud (2009) found that only 34% of trials reported all data required to calculate the sample size, had accurate calculation and used accurate assumptions for the control group. Descôteaux (2007) concludes that the new awareness of power exists mostly in theory as the power levels of studies differed little from the 48% reported by Cohen (1962).

1.5 Aim and contents of the book

1.5.1 Aim

Power analyses have gained the attention of applied statisticians in the fields of social and behavioral sciences and medicine. Researchers planning a study may consult many handbooks on power calculations, such as those by Bausell and Li (2002); Chow, Shao, and Wang (2008); Cohen (1988); Dattalo (2008); Julious (2010); Liu (2014); Machin, Campbell, Tan, and Tan (2009); Murphy, Myors, and Wolach (2008); Ryan (2013), to name but a few. In addition, computer programs for statistical power analysis are available, such as G*Power (Faul, Erdfelder, Lang, & Buchner, 2007; Mayr, Erdfelder, Buchner, & Faul, 2007), nQuery Advisor (Elashoff, 2007), and PASS (Hintze, 2008).

These books and software packages mainly apply to trials without a hierarchical data structure. Much attention has been paid to sample sizes for trials with hierarchical data in the 1990s and 2000s and sample size formulae have appeared in journals in the social and biomedical fields. Until now, sample size guidelines for such trials were scattered over chapters in existing books. Introductions to sample sizes for trials with nested data in general can be found in Berger and Wong (2009), H. Brown and Prescott (2015), J. J. Hox (2010), Snijders (2001) and Snijders and Bosker (2012). Sample sizes for cluster randomized trials are presented in M. J. Campbell and Walters (2014), Donner and Klar (2000b), S. Eldridge and Kerry (2012), Hayes and Moulton (2009), Moerbeek, Van Breukelen, and Berger (2008), Moerbeek (2006a), Moerbeek and Teerenstra (2010), Murray (1998) and Van Breukelen (2013).

The aim of this book is to summarize methods for sample size calculations for various types of trials with hierarchial data, thereby providing social and biomedical researchers a useful handbook that guides them in their choice of the best research design and the required sample sizes.

1.5.2 Contents

This book consists of 12 chapters. To use the sample size formulae, a basic understanding of statistical models for the analysis of trials with hierarchial data is desirable. An introduction to these models is given in Chapter 2. Chapter 3 gives an overview of the concepts of statistical power analysis, the methods that are available to perform a power calculation and the steps that should be taken in a power analysis. The reader is advised to first to read Chapters 2 and 3 to become familiar with terminology and notation before proceeding to the subsequent chapters.

Chapter 4 covers sample sizes for cluster randomized trials. Sample size formulae are provided when the number of clusters or the cluster size is fixed beforehand, or when both are undetermined and the researcher wishes to minimize the costs for including clusters and subjects within clusters. Sample size guidelines are provided for continuous and binary outcome variables.

Cluster randomized trials are often hampered by low statistical power as compared to simple randomized trials, especially when the intraclass correlation coefficient and/or group size is large. Methods to improve statistical power in cluster randomized trials are provided in Chapter 5: including covariates, taking repeat measurements and using a crossover design.

Covariates are variables at the cluster or subject level that are related to the outcome variable. Including covariates often results in lower error variances and hence a higher power level, provided that such covariates are uncorrelated with treatment condition. This condition may be fulfilled by random assignment of clusters to treatment conditions when the number of clusters is large, and by minimization, matching or pre-stratification when the number of clusters is small.

Another method to improve power is to include the pre-test score on the outcome variable in a design where repeat measurements are taken. With a crossover design clusters are randomized to sequences of treatment conditions such that multiple treatments are available within each cluster. The stepped-wedge design can be considered a special case of the crossover design where the intervention is sequentially rolled out to clusters over a number of time periods.

Chapters 6, 7 and 8 focus on multisite trials, pseudo cluster randomized trials and individually randomized group treatment trials, respectively. These chapters share a common structure: the first section gives an introduction to the design and refers to relevant literature, including special issues of international peer-reviewed journals. As it is futile to provide sample size formulae and discuss design optimality without reference to the appropriate analysis model, the second sections of these chapters give the multilevel model. For cluster randomized trials and multisite trials, these models are related to the analysis of variance models that are well known in many disciplines of the social and biomedical sciences. The third and fourth sections of each chapter focus on power analyses for continuous and dichotomous outcomes. The final section of each chapter gives an example to illustrate the use of the sample size formulae provided in that chapter.

Sample sizes for longitudinal intervention studies are provided in Chapter 9. The power for such trials depends on the number of subjects and repeated measurement per subject and also on the duration of the study. As subjects' drop-outs from the trial increase with increasing study duration, it is important to study the effects of different drop-out patterns on the power level.

Chapter 10 provides sample size calculations for other designs with subjects nested within groups. It deals with cluster randomized trials with three levels of nesting and with multisite cluster randomized trials, where cluster randomization is performed within multiple sites. It also provides sample size formulae for cluster randomized trials and multisite trials with repeated measurements. In addition, it deals with factorial designs where more than two treatments are compared.

The true value of the intraclass correlation coefficient needs to be known to calculate the required sample size in a cluster randomized trial. This value is generally unknown and methods to overcome this problem are provided in Chapter 11. Sample size re-estimation uses an estimate of the intraclass correlation coefficient as obtained from an internal pilot to estimate the required sample size for the remainder of the study. Bayesian optimal designs use prior model parameters to calculate an interval of the required sample size for which the desired power level is achieved. In addition, this chapter provides an overview of estimates of intraclass correlation coefficients from previous trials as reported in the literature.

The last chapter gives an overview of existing and new software to perform power calculations. A freeware computer program to perform the power calculations in this book has been developed by the authors, and a basic introduction to this program is provided in the last chapter.

2

Multilevel statistical models

Models to analyze hierarchical data structures are known under a variety of names. In the social and behavioral sciences, they are often referred to as multilevel models, hierarchical (linear) models, random coefficient modelsand variance component models. In the biomedical sciences, mixed and random (effects) models are more common terms. In the remainder of this book the term multilevel model will be used.

The aim of this chapter is to provide a short introduction to multilevel models for continuous and dichotomous outcomes. For a more exhaustive introduction the reader is referred to the handbooks by Gelman and Hill (2007), Goldstein (2011), Heck and Thomas (2015), J. J. Hox (2010), Longford (1993), Raudenbush and Bryk (2002), and Snijders and Bosker (2012). Hedeker and Gibbons (2006) provide a good overview of multilevel models for ordinal and nominal dependent variables in the context of longitudinal data analysis. Singer and Willett (2003) provide an introduction to linear and nonlinear multilevel models for longitudinal data analysis but restrict the discussion to continuous outcome variables. A glossary of terms that are common in the field of multilevel analysis is provided by Diez Roux (2002).

2.1 The basic two-level model

The simplest form of the multilevel model has two levels of nesting. In general, the units at these levels are referred to as subjects and groups or clusters. These terms may be replaced by appropriate terms from the reader's field of science. In educational sciences, for instance, pupils are nested within schools; in family studies, family members are nested within families; and in clinical trials, primary care patients are nested within primary care practices.

The two-level model formulates separate models for both levels of the hierarchical data structure that are then combined into a single equation model. The outcome variable is measured on the subject level. The effects of predictor variables at the subject level on the outcome variable are allowed to vary across the groups in which the subjects are nested. So, a certain predictor variable may have a large effect on the outcome in the one group whereas its effect in the other group may be negligible. It may also occur that a positive relation between predictor and outcome exists in some groups while this relation is negative in others. This variation in the relation between predictor and outcome may subsequently be modeled as a function of predictor variables at the group level.

In the following model formulation, a model with a single predictor variable at both the subject and group levels is assumed. The continuous outcome variable y_{ij} of subject i in group j is predicted from the predictor x_{ij} at the subject level:

$$y_{ij} = \beta_{0j} + \beta_{1j}x_{ij} + e_{ij}. \tag{2.1}$$

Model (2.1) is called the level-one model or subject level model, since it does not contain any predictor variables at the group level. Notice that both the response y_{ij} and predictor x_{ij} have subscripts i and j since they vary at both the subject and group level. The residual e_{ij} of subject i in group j represents the deviation between this individual's observed response y_{ij} and the expected response $\beta_{0j} + \beta_{1j}x_{ij}$ in that group, given x_{ij}. The residuals e_{ij} at the subject level are assumed to be independently and normally distributed with zero mean and variance σ_e^2.

The intercept β_{0j} and slope β_{1j} are called random effects since they are allowed to vary over the groups, as indicated by their subscripts j. They can be split into an average component (represented by γ) and a group-dependent deviation (represented by u):

$$\begin{aligned} \beta_{0j} &= \gamma_{00} + u_{0j} \\ \beta_{1j} &= \gamma_{10} + u_{1j}. \end{aligned} \tag{2.2}$$

Model (2.2) is called the unconditional level-two model or the group level model. It should be noted that the effects γ are not subscripted j. In other words, they are constant across groups and are therefore called fixed effects. The term γ_{00} is the average intercept and represents the mean score when the predictor variable $x_{ij} = 0$. The term γ_{10} is the average slope and represents the mean increase in outcome y_{ij} when the predictor x_{ij} increases one unit. The residuals u_{0j} and u_{1j} at the group level represent the deviation of group j from the average intercept and slope. The pairs (u_{0j}, u_{1j}) are assumed independent and follow a bivariate normal distribution with zero means, variances σ_{u0}^2 and σ_{u1}^2, and covariance σ_{u01}. In addition, they are assumed to be independent of the level-one residuals e_{ij}.

The covariance represents the degree to which the intercepts and slopes covary. A positive covariance implies that a higher score at $x_{ij} = 0$ is related to a higher effect of the predictor x_{ij}, whereas a negative covariance implies that a higher intercept is related to a lower effect of the predictor.

Substitution of the level-two model (2.2) into the level-one model (2.1) results in the single-equation multilevel model

$$y_{ij} = \gamma_{00} + \gamma_{10}x_{ij} + u_{0j} + x_{ij}u_{1j} + e_{ij}. \tag{2.3}$$

The first part of the model, $\gamma_{00} + \gamma_{10}x_{ij}$, contains the fixed coefficients γ and represents the average relation between the predictor x_{ij} and response y_{ij}. It is often referred to as the fixed or deterministic part. The second part, $u_{0j} + x_{ij}u_{1j} + e_{ij}$, is a function of the random error terms and is therefore referred to as the random or stochastic part. As the model contains fixed and random effects, it is referred to as a mixed (effects) model.

An example: The relation between socioeconomic status and adolescent depression. Consider as a hypothetical example a study that aims to determine whether adolescent depression is related to socioeconomic status (SES) and whether this relationship depends on the school context. In this study, adolescents are nested within schools and the level-one model that describes the relationship between the continuous outcome variable depression and predictor variable SES is given by

$$depression_{ij} = \beta_{0j} + \beta_{1j}SES_{ij} + e_{ij}. \tag{2.4}$$

As the intercept and slope vary across schools, each school has its own relationship between SES and depression. This is shown in Figure 2.1, in which this relationship is depicted for a sample from a population of schools. Each school is represented by a regression line; the bold line represents the mean relationship in the population. Figure 2.1 shows that on average the relationship between SES and depression is negative: the higher SES, the lower the degree of depression. Indeed, most lines show a decreasing relation of depression as a function of SES. However, there are some schools for which the relationship is positive, as is shown by their increasing lines. For these schools, high SES is related to high degrees of depression.

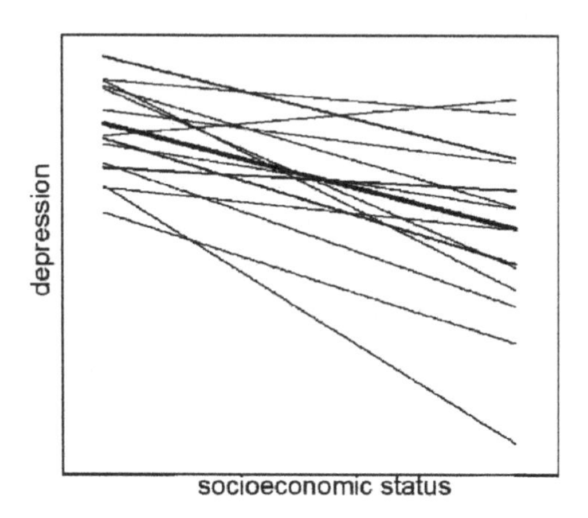

Figure 2.1: The relation between SES and adolescent depression for a sample of schools. The bold line represents the mean relationship.

The intercept for group j has a normal distribution with mean γ_{00} and variance σ^2_{u0}. Consequently, 95% of the groups have an intercept above $\gamma_{00} - 1.96\sigma_{u0}$ and below $\gamma_{00}+1.96\sigma_{u0}$, where the standard deviation σ_{u0} of the random effect

u_{0j} is the square root of the variance σ^2_{u0}. A similar interval can be constructed for the slopes. The higher the variances σ^2_{u0} and σ^2_{u1}, the higher the between-group variation with respect to intercept and slope, and the broader their intervals. Part of this variation can be explained by a group level predictor variable z_j:

$$\begin{aligned} \beta_{0j} &= \gamma_{00} + \gamma_{01} z_j + u_{0j} \\ \beta_{1j} &= \gamma_{10} + \gamma_{11} z_j + u_{1j}. \end{aligned} \tag{2.5}$$

As the predictor variable z_j varies over groups but not over subjects within groups, it is indicated by j instead of ij. The residuals u_{0j} and u_{1j} in the conditional group level model (2.5) represent the sources of variation in the intercept β_{0j} and slope β_{1j} that are not explained by the predictor variable z_j. It is expected that their variances σ^2_{u0} and σ^2_{u1} are lower than those of their counterparts in the unconditional model (2.2) as the predictor z_j explains part of the variance in intercept and slope.

Substitution of the level-two model (2.5) into the level-one model (2.1) results in the single-equation or combined multilevel model

$$y_{ij} = \gamma_{00} + \gamma_{01} z_j + \gamma_{10} x_{ij} + \gamma_{11} x_{ij} z_j + u_{0j} + u_{1j} x_{ij} + e_{ij}. \tag{2.6}$$

This model contains an interaction term $x_{ij} z_j$ that consists of a predictor x_{ij} at the subject level and a predictor z_j at the group level. Such an interaction term is called a cross-level interaction. It implies that the effect of the subject level predictor variable x_{ij} is moderated by the predictor z_j at the group level. The size of the interaction effect is denoted γ_{11}. In addition, the model also includes main effects γ_{01} and γ_{10} of the predictors z_j and x_{ij}.

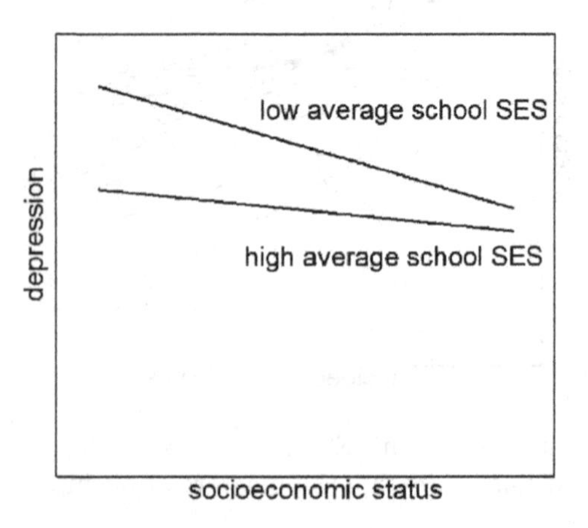

Figure 2.2: The relation between SES and adolescent depression for high and low values of school average SES.

Model (2.6) is called a linear model since it is a linear function of the regression coefficients γ. A linear regression model is not necessarily linear with respect to the predictor variables. A model that contains the square of a predictor, for instance, is still considered to be linear.

An example: The relation between socioeconomic status and adolescent depression. In the example on Page 23, low levels of SES were related to high levels of adolescent depression. To examine whether the school context has an effect on this relationship, a predictor variable at the school level is included in the model. A relevant predictor variable is a school's average SES. This is an aggregated variable since it aggregates (i.e., averages) a subject level variable to the group level. The multilevel model now becomes

$$\begin{aligned} depression_{ij} &= \gamma_{00} + \gamma_{01}\overline{SES}_j + \gamma_{10}SES_{ij} + \gamma_{11}SES_{ij}\overline{SES}_j \\ &\quad + u_{0j} + u_{1j}SES_{ij} + e_{ij}, \end{aligned} \tag{2.7}$$

where $\overline{SES}_j$ denotes the average SES of adolescents in school j. The multilevel model shows an interaction effect between SES and its school average. Figure 2.2 shows the moderating effect of average school SES on the relation between an adolescent's SES and depression. The two bold lines represent the average relation for schools at the lower and upper quartile of average school SES. For clarity reasons, the lines representing the separate schools are not shown in this graph. On average, adolescent depression at a given value of SES is lower in schools with a higher average SES. The relation between SES and adolescent depression is negative for both low and high average SES schools, although it is somewhat stronger in schools with a low average school SES.

The example shows that the interpretation of main and interaction effects in the fixed part of the model is similar to that in a traditional regression model. However, it should be understood that each school may deviate from these mean relationships due to the model's random part.

The example used average school SES as a school level predictor variable. Other relevant predictors at the school level might be the percentage of minorities, school type (public versus private) and location (metropolitan versus rural). Relevant additional predictors at the adolescent level might be gender and age. The multilevel model (2.6) may be extended to any number of predictor variables at the subject and group level. Suppose the model includes P predictors at the subject level, and each of them has a random effect. The subject level model is then represented as

$$y_{ij} = \beta_{0j} + \beta_{1j}x_{1ij} + \ldots + \beta_{Pj}x_{Pij} + e_{ij}. \tag{2.8}$$

Each random effect β_{pj} is estimated from Q predictor variables at the group level:

$$\beta_{pj} = \gamma_{p0} + \gamma_{p1}z_{1j} + \ldots + \gamma_{pQ}z_{Qj} + u_{pj}. \tag{2.9}$$

The single level equation is obtained by substitution of (2.9) into (2.8):

$$\begin{aligned} y_{ij} &= \gamma_{00} + \sum_{p=1}^{P} \gamma_{p0} x_{pij} + \sum_{q=1}^{Q} \gamma_{0q} z_{qj} + \sum_{p=1}^{P} \sum_{q=1}^{Q} \gamma_{pq} z_{qj} x_{pij} \\ &\quad + u_{0j} + \sum_{p=1}^{P} u_{pj} x_{pij} + e_{ij}. \end{aligned} \tag{2.10}$$

Observe that group level variables that do not explain (partially) the random variation of a subject level predictor, but are directly related to the outcome (e.g., school type), are included in the model as group-level predictor variables of β_{0j}. Again, the model consists of a fixed part (right side of first line) and a random part (second line). The random effects u_{pj} at the group level are assumed to be independent of the subject level residuals e_{ij}, which themselves are assumed independent. The variances of the u_{pj} are $var(u_{pj}) = \sigma^2_{up}$ and they have covariances $cov(u_{pj}, u_{p'j}) = \sigma_{upp'}$ $(p \neq p')$.

Even for a small number of predictors at the subject and group level the multilevel model (2.10) may become very complicated. In practice, the model is simplified by removing non-significant fixed effects and random effects for which the variance is not significantly larger than zero. Hypothesis testing is discussed in the next section.

2.2 Estimation and hypothesis test

This section gives a brief overview of methods for parameter estimation and hypothesis testing in multilevel models. A more extensive overview can be found in the handbooks that are referred to at the beginning of this chapter. It should be mentioned that a thorough understanding of estimating and testing methods is not necessary to understand the power calculations in the next chapters.

As it contains four predictor variables, the multilevel model (2.6) can be viewed upon as a multiple linear regression model. It differs from a traditional multiple linear regression model, however, since it includes random effects u_{0j} and u_{1j} at the group level. These random effects are necessary to take into account the nested structure of the data. As a consequence, the variance of a subject's outcome y_{ij}, conditional on its subject level predictor variable x_{ij}, depends on the between-group variability:

$$var(y_{ij}|x_{ij}) = \sigma^2_{u0} + 2\sigma_{u01} x_{ij} + \sigma^2_{u1} x^2_{ij} + \sigma^2_e. \tag{2.11}$$

It is observed that this variance is a function of x_{ij}; hence the multilevel model allows for heteroscedasticity. The covariance of two subjects i and i' $(i \neq i')$

within the same group is equal to

$$cov(y_{ij}, y_{i'j}|x_{ij}, x_{i'j}) = \sigma^2_{u0} + \sigma_{u01}(x_{ij} + x_{i'j}) + \sigma^2_{u1}x_{ij}x_{i'j}. \tag{2.12}$$

This covariance is generally unequal to zero, thus the multilevel takes into account the dependency of outcomes of subjects within the same group. The covariance of outcomes of subjects within different groups j and j' ($j \neq j'$) is equal to zero:

$$cov(y_{ij}, y_{i'j'}|x_{ij}, x_{i'j'}) = 0. \tag{2.13}$$

Thus, the multilevel model assumes that outcomes of subjects within different groups are independent.

The multilevel model (2.6) allows for heteroscedasticity and dependent outcomes within the same group. This also applies to models with more predictor variables at the subject level, where the variance of an outcome and the covariance of outcomes within the same group is a function of all subject level predictor variables. Traditional estimation techniques such as ordinary least squares (OLS) should therefore not be used to estimate model parameters in a multilevel model since they will result in incorrect point estimates and corresponding standard errors. As a consequence, tests with respect to the effects of predictor variables will be too liberal or too conservative, depending on the level at which the predictor variables are measured and the sizes of the variances σ^2_{up} (Moerbeek et al., 2003a).

Regression coefficients and (co)variances in multilevel models are generally estimated by estimation methods such as full information maximum likelihood (ML) or restricted maximum likelihood (REML). Maximum likelihood methods estimate the model parameters (i.e., regression coefficients and (co)variances) such that the likelihood function is maximized. The likelihood is the probability of getting the sample data, given the statistical model and the values of the model parameters.

The ML and REML methods differ little with respect to their regression coefficient estimates, but they do differ with respect to their variance estimates. This is explained by the fact that the REML method maximizes the part of the likelihood function that is invariant to the regression coefficients, while the ML method maximizes the likelihood as a function of regression coefficients and (co)variances. In other words, the REML method takes into account the loss of degrees of freedom that arises from the estimation of the fixed effects. Thus, REML estimates of the variances are unbiased, while those obtained with ML have a downward bias. This bias is only substantial if the sample size is small and in this case the REML method is preferred. For large sample sizes (say, more than 30 groups), the difference between ML and REML variance estimates becomes negligible.

ML and REML estimation methods were proposed in the 1960s and 1970s (Hartley & Rao, 1967; W. A. Thompson, 1962). During the 1980s much research was done on the development of algorithms that compute ML and REML estimates, such as iterative generalized least squares (IGLS) and restricted iterative generalized least squares (RIGLS) (Goldstein, 1986, 1989). As indicated by their names, these algorithms require an iterative procedure.

Based on some starting values, which are typically obtained from a single-level analysis, the regression coefficients are estimated, and their estimates are subsequently used to estimate the (co)variances. These (co)variances are then used to estimate the regression coefficients, etcetera. These steps are iterated many times and in each step the estimates are compared to the estimates from the previous iteration. If the difference between estimates from successive iterations is sufficiently small, the maximum value of the likelihood is reached. This means that convergence is achieved and the estimation process terminates.

In addition to the development of new algorithms, existing algorithms were adopted to multilevel models, such as the Fisher scoring algorithm (Longford, 1987) and the expectation-maximization (EM) algorithm (Longford, 1987) (Dempster, Rubin, & Tsutakawa, 1981; Mason, Wong, & Entwisle, 1983). In the end these methods provide the same parameters for models with continuous outcomes, except from some small differences arising from the chosen convergence criterion.

It should be noted that the random effects u_{pj} are treated as latent parameters by the multilevel model (2.10). They are therefore not part of the likelihood function. Instead, their variances and covariances are estimated to provide a measure of the between-group variability of intercepts and parameter effects. In some studies it is useful to have estimates of these random effects, for instance when the groups in the data set are to be compared, such as is the case with school effectiveness studies. The random effects can be estimated with empirical Bayes estimation. As the focus of the multilevel models in the succeeding chapters is on the treatment effect, which is a parameter in the fixed part of the model, the estimation of random effects is not further discussed. Details of empirical Bayes estimation can be found in the books that were cited in the beginning of this chapter.

Maximum likelihood methods do not only produce point estimates of regression coefficients and (co)variances of random effects, but also their associated standard errors. These are asymptotic estimates since they are only valid for large samples. Point estimates and standard errors of fixed effects can be used to calculate the Wald test statistic $z = (point\ estimate)/(standard\ error)$. For large sample sizes, this test statistic follows a standard normal distribution under the null hypothesis of no effect.

The Wald test is generally discouraged for testing variances and covariances, especially when the sample size is small. For such parameters, procedures based upon the deviance test statistic are preferred. The deviance is a statistic that is calculated as minus twice the natural logarithm of the maximum likelihood value, that is, minus twice the log likelihood evaluated at the estimated values of the model parameters. It provides an indication of lack of fit between the observed data and the model. Lower deviances therefore correspond to better fit, but the value of the deviance cannot be interpreted directly.

Deviance statistics can be used to compare two nested models that are fitted to the same data set. Two models are considered nested if all parameters in the simpler model also appear in the more complex model. Thus, the more complex model is an extension to the simpler model as its adds one or more regression weights or random effects. Two nested models can be compared by calculating the difference of their deviances and referring it to the chi-squared distribution with degrees of freedom equal to the difference in number of parameters between the two models. The simpler model corresponds to the null hypothesis that the added parameters are all equal to zero.

For instance, to test whether an intercept in a two level model has a random effect, one should fit a model with a fixed intercept and another model with a random intercept and subtract their deviances. Under the null hypothesis of a fixed intercept, this statistic follows a chi-squared distribution with one degree of freedom, as the variance of the intercept is the sole parameter that is added to the simpler model. The p-value of the test is equal to the upper tail probability of the value of the test statistic divided by two. A division by two is made since variance parameters are non-negative by definition. Thus, the corresponding alternative hypothesis is one-sided (Berkhof & Snijders, 2001). Such a division should not be made if the deviance test is used for testing fixed parameters, unless one has a one-sided alternative hypothesis in mind.

The ML estimation method should be used when fixed parameters are tested by means of a deviance test. In case two models are compared that differ with respect to their random part, but not to their fixed part, then the deviances as produced by the REML method can also be used in a deviance test.

2.3 Intraclass correlation coefficient

As explained previously, the multilevel model takes account of the dependency of outcomes within the same group. The degree of this dependency is expressed by means of a statistic that is usually referred to as the intraclass correlation coefficient (ICC). As it does not only apply when subjects are nested in school classes, but in any unit at the second level, the terms intra-level-two-unit correlation coefficient, intra-group correlation coefficient and intra-cluster correlation coefficient are more generally appropriate. The ICC is calculated on the basis of a multilevel model without any predictor variables. Such a model is called an intercept-only model or an empty model and the formulation is

$$y_{ij} = \gamma_{00} + u_{0j} + e_{ij}. \tag{2.14}$$

The fixed effect γ_{00} represents the overall mean and the terms u_{0j} and e_{ij} are the residuals at the group and subject level that have zero means and variances σ_{u0}^2 and σ_e^2. The variance of subject i in group j is

$$var(y_{ij}) = \sigma_{u0}^2 + \sigma_e^2. \tag{2.15}$$

This variance consists of two components σ_{u0}^2 and σ_e^2, which are therefore called variance components. The covariance of two observations i and i' $(i \neq i')$ in the same group j is equal to

$$cov(y_{ij}, y_{i'j}) = \sigma_{u0}^2. \tag{2.16}$$

The intraclass correlation coefficient ρ is defined as

$$\rho = \frac{cov(y_{ij}, y_{i'j})}{var(y_{ij})} = \frac{\sigma_{u0}^2}{\sigma_{u0}^2 + \sigma_e^2}, \tag{2.17}$$

and is a measure of the proportion variance of the outcome y_{ij} that is located at the group level. It is interpreted as the expected correlation between outcomes of two randomly chosen subjects within the same group.

For the empty model, closed form estimators for the variance components exist:

$$\hat{\sigma}_e^2 = MS(Error), \tag{2.18}$$

and

$$\hat{\sigma}_{u0}^2 = \frac{MS(Group) - MS(Error)}{(N - \sum_j n_{1j}^2/N)/(n_2 - 1)}, \tag{2.19}$$

see for instance Searle, Casella, and McCulloch (1992). In some cases the estimate $\hat{\sigma}_{u0}^2$ is negative and is then replaced by the value zero. n_2 is the number of groups, n_{1j} is the number of subjects in group j, and $N = \sum_j n_{1j}$ is the total sample size. The mean squares $MS(Error)$ and $MS(Group)$ are calculated as follows:

$$MS(Error) = \frac{\sum_j \sum_i (y_{ij} - \bar{y}_{.j})^2}{N - n_2} \tag{2.20}$$

and

$$MS(Group) = \frac{\sum_j n_j (\bar{y}_{.j} - \bar{y}_{..})^2}{n_2 - 1}, \tag{2.21}$$

where $\bar{y}_{..}$ is the overall mean response and $\bar{y}_{.j}$ is the mean response in group j. The estimates (2.18) and (2.19) can be substituted in Equation (2.17) to obtain an estimate of the intraclass correlation coefficient. For more complicated models, no simple closed form expressions for the variance components estimates exist. For such models, we must rely on iterative estimation methods described in the previous section.

The ICC is non-negative and lies in the interval [0, 1]. The higher its value, the higher the degree of correlation of outcomes of subjects within the same group. Values close to one are rarely encountered in practical situations. The ICC tends to be inversely related to cluster size: the higher the cluster size, the lower the ICC and vice versa (Gulliford et al., 2005). ICC values in general practices with hundreds or even thousands of patients, for instance, tend to be very small, while ICC values in small units, such as families, are generally much higher. This may be explained by the fact that the degree of communication and mutual influence between subjects in a large level-two unit is in general much lower than that in a small level-two unit. In addition, the degree of dependency of family members in the same family is often relatively large due to genetic similarities between family members.

The ICC has an effect on the degree to which incorrect conclusions are achieved when the multilevel data structure is ignored and a traditional unilevel regression model is used. Consider as an example the estimation of the mean of a particular variable. The estimated mean is simply equal to the average of the observations in the sample. If one ignores the nesting of subjects in groups, the corresponding variance is calculated as σ^2/N, where σ^2 is the between-subject variance and N is the sample size. When taking the hierarchical data structure into account, this variance is inflated by the factor $1 + (n_1 - 1)\rho$, where n_1 is the common group size. For varying group sizes, this factor is approximated by replacing n_1 by the average group size $\overline{n}_1$. The factor $1 + (n_1 - 1)\rho$ is the so-called design effect, and is well-known in survey research (Kish, 1965), where it determines the effect of taking a cluster sample instead of a simple random sample.

Ignoring the multilevel data structure does not have an effect on the variance of the mean if the design effect is equal to one. This occurs if outcomes of subjects within the same group are uncorrelated (i.e., if $\rho = 0$) or if just one subject is sampled within each group (i.e., if $n_1 = 0$). For all other situations, the design effect is larger than one, resulting in an underestimated variance if a traditional regression model is used. Even for small values of the ICC, the design effect may be considerable. For instance, it is equal to 1.95 if the group size is equal to 20 and $\rho = 0.05$, meaning that the variance of the mean as obtained with a multilevel model is almost twice as large as that from a traditional regression model. Figure 2.3 shows the design effect as a function of ICC for three different group sizes. For any group size it is equal to 1 when $\rho = 0$. For all $\rho > 0$, ignoring the multilevel data structure results in an underestimated variance of the mean, and hence too liberal statistical tests.

It should be noted that each of the designs covered in this book has its own design effect. Such a design effect is not always larger than one, nor does it always depend on the group size. The chapter on multisite experiments includes a situation where the design effect for the treatment effect is independent of group size and less than one. In such a situation, ignoring the multilevel data structure results in a too conservative test with respect to the effect of treatment.

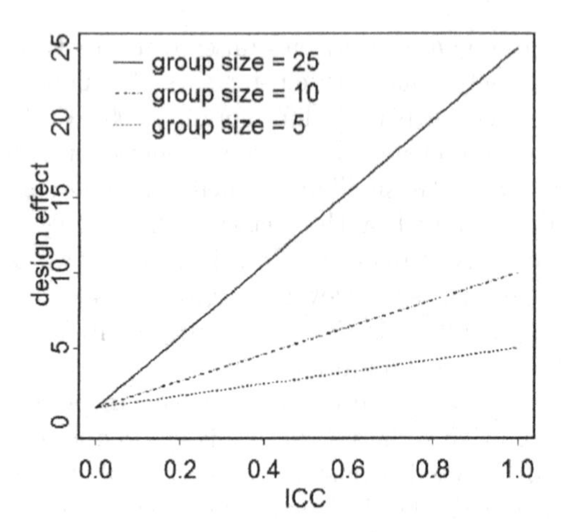

Figure 2.3: Design effect as a function of ICC and group size.

2.4 Multilevel models for dichotomous outcomes

The multilevel models discussed thus far were restricted to continuous outcome variables. In many real-life situations, outcomes are measured on a dichotomous or binary scale, meaning that just two possible outcome values are possible. For instance, an adolescent within a particular school may smoke or not, a patient within an oncologic unit may have recovered from cancer or not, or an employee within a firm may be on sick leave or not.

Multilevel models for non-continuous outcomes are extensions to generalized linear models (McCullagh & Nelder, 1989). They are referred to as generalized linear mixed models or hierarchical generalized linear models and take into account the nesting of subjects within groups by allowing the effects of subject level predictors to randomly vary over groups.

The commonly used formulation of the multilevel model for a binary response y_{ij} of subject i in group j is

$$y_{ij} = p_{ij} + e_{ij}, \tag{2.22}$$

where p_{ij} is the probability of a response $y_{ij} = 1$ and the residual e_{ij} is the discrepancy between the observed response y_{ij} and the probability p_{ij}. This residual has zero mean and variance $p_{ij}(1-p_{ij})$, since the outcome y_{ij} follows a Bernoulli distribution.

The probability p_{ij} for subject i in group j is related to the linear predictor η_{ij}, which consists of the predictor variables x_{ij} and z_j and their fixed and/or random effects. For instance, $\eta_{ij} = \gamma_{00} + \gamma_{10}x_{ij} + u_{0j} + u_{1j}x_{ij}$ for the model with a random intercept and a single subject level predictor x_{ij} that has a random effect. The relation between p_{ij} and η_{ij} is described by means of a link function g:

$$\eta_{ij} = g(p_{ij}) \text{ and } p_{ij} = g^{-1}(\eta_{ij}), \tag{2.23}$$

where g^{-1} is the inverse link function. The link function g should be chosen in such a way that the predicted probability is in the interval $[0, 1]$. Thus, the linear link function $\eta_{ij} = p_{ij}$ is not a good choice as the probability may fall outside this interval for large or small values of the predictor variables. Various link functions have been proposed, of which the logit, probit and complementary log-log functions are the most common. In this section the focus is on the logit link function, which is defined as

$$logit(p_{ij}) = ln(odds(p_{ij})) = ln\left(\frac{p_{ij}}{1 - p_{ij}}\right) = \eta_{ij}. \tag{2.24}$$

The linear predictor η_{ij} is the natural logarithm of the odds $p_{ij}/(1 - p_{ij})$. Rearranging this relationship results in

$$p_{ij} = \frac{1}{1 + e^{-\eta_{ij}}}. \tag{2.25}$$

Alternatively, the multilevel logistic model may be represented using a threshold model, which assumes that a continuous latent variable $\widetilde{y}_{ij} = \eta_{ij} + \epsilon_{ij}$ underlies the dichotomous response y_{ij}. The error ϵ is assumed to follow the standard logistic distribution with zero mean and variance $\pi^2/3$. An outcome $y_{ij} = 1$ occurs if the latent variable $\widetilde{y}$ exceeds a threshold, otherwise the outcome is $y_{ij} = 0$. This threshold is commonly fixed to be equal to zero.

It should be noted that each group j may have its own relationship between the response probability p_{ij} and the predictor variables. This is illustrated in Figure 2.4 where the relation between a pupil's perceived group pressure and the probability of adolescent smoking is depicted for a sample of schools. The model is supposed to have a random intercept and random effect of perceived group pressure, but no predictor variables at the school level are included. Thus, the linear predictor is equal to $\gamma_{00} + \gamma_{10} group\ pressure_{ij} + u_{0j} + u_{1j} group\ pressure_{ij}$. The bold line shows that on average the probability of adolescent smoking increases with increasing perceived group pressure. If perceived group pressure is absent, then the average probability to start smoking is about 0.2. The thin lines show the relationships for a sample of schools.

If a school's random effect u_{0j} is positive, then the probability of adolescent smoking at zero perceived group pressure is larger than the average probability (i.e., the schools's line intersects the vertical axis at a higher value). If a school's random effect u_{1j} is positive, then the relationship between perceived group pressure and the probability of adolescent smoking is stronger than the average relation. This is depicted by the school's line having a steeper slope than the bold line. For one school in Figure 2.4 the relationship is negative, resulting from a large negative value u_{1j}.

The intraclass correlation coefficient of the multilevel logistic model as defined

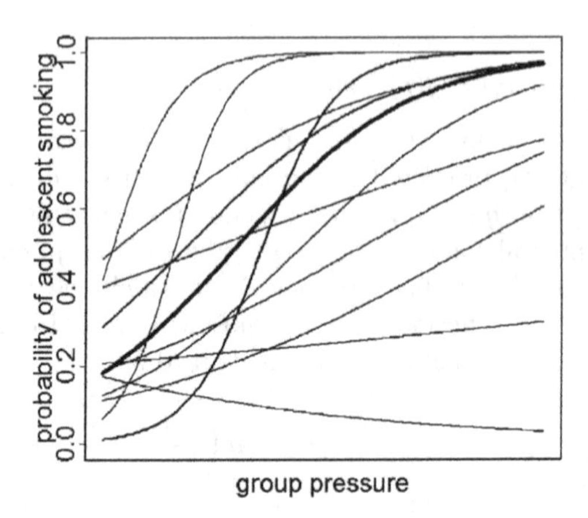

Figure 2.4: The relation between group pressure and the probability of adolescent smoking for a sample of schools. The bold line represents the mean relationship.

by Equations (2.22) and (2.25) can be calculated in different ways, of which two are shortly discussed here. The first uses the mean squares in Equations (2.18) and (2.19) with the continuous outcome y_{ij} replaced by the estimated proportion $\hat{p}_{ij}$ to calculate the variance components, which are then imputed in (2.17) to calculate the intraclass correlation coefficient. The second uses the fact that the individual level error term ϵ_{ij} in the threshold representation of the model has a variance $\pi^2/3$. Replacing σ_e^2 in Equation (2.17) by $\pi^2/3$ results in $\rho = \sigma_{u0}^2/(\sigma_{u0}^2 + \pi^2/3)$.

Estimating model parameters in a generalized linear mixed model is more complicated than estimating parameters in a linear mixed model. This is caused by the fact that the likelihood function contains an integral that cannot be solved analytically. In the past decades various approximation methods have been proposed, such as marginal and penalized quasi-likelihood (MQL and PQL) that rely on Taylor series expansions (Goldstein, 1991; Goldstein & Rasbash, 1996). These methods include either a first order term or first and second order terms of the Taylor series expansion. First order MQL offers the crudest approximation and is discouraged (Rodríguez & Goldman, 1995). Second order PQL is a better approximation, although the fixed and random effects of the model may be substantially underestimated when the random effects are sizable (Rodríguez & Goldman, 2001).

Moerbeek, Van Breukelen, and Berger (2003b) performed a simulation study to evaluate the performance of first and second order PQL in the context of cluster randomized trials and multisite trials. They focused on the parameter that represents the treatment effect, as this is the parameter of main interest in experiments. They concluded that the estimates of this parameter as obtained with second order PQL seldom deviated more that 5% from the true value and that the standard errors were almost unbiased. For this estimation method the standard normal approximation may be used as a reference distribution for testing the treatment effect. PQL approximation is preferred above MQL approximation. The treatment effect can be tested by means of the Wald test, with test statistic $z = (point\ estimate)/(standard\ error)$.

Other approaches use a combination of Taylor series expansion and Laplace approximation (Raudenbush, Yang, & Yosef, 2000), or numeric integration of the likelihood (Hedeker & Gibbons, 1994). All these methods may result in slightly different parameter estimates, and under extreme cases where complex models are fitted to small data sets the estimation methods may fail to achieve convergence or run into numerical problems. It is therefore good practice to compare parameter estimates for generalized linear mixed models as obtained with various estimation methods and to check whether convergence has indeed been achieved.

2.5 More than two levels of nesting

The linear two-level models discussed so far can be easily extended by including additional levels of nesting. In educational sciences, for instance, pupils are nested within classes within schools, patients in health care studies are nested within physicians within clinics and employees in organizational studies employees are nested within units within companies. As ignoring levels in the multilevel data structure may result in incorrect conclusions with respect to model parameters and their standard errors, it is important to carefully identify all levels in the data structure (Berkhof & Kampen, 2004; Moerbeek, 2004; Opdenakker & Van Damme, 2000; Tranmer & Steel, 2001; Van Den Noortgate, Opdenakker, & Onghena, 2005).

The formulation of models with more than two levels of nesting is similar to that of two-level models. The outcome variable is measured at the individual level. Both the intercept and the effects of subject level predictor variables are allowed to vary over the groups and all higher level units. Here a three-level model with pupils in classes in schools is considered. The level-one model of pupil i in class j in school k with one predictor at the pupil level is given by

$$y_{ijk} = \pi_{0jk} + \pi_{1jk}x_{1ijk} + e_{ijk}. \tag{2.26}$$

The intercept π_{0jk} and slope π_{1jk} are subscripted jk to indicate that they vary over the classes and schools. The class level model is given by

$$\begin{aligned} \pi_{0jk} &= \beta_{00k} + u_{0jk} \\ \pi_{1jk} &= \beta_{10k} + u_{1jk}. \end{aligned} \tag{2.27}$$

The regression coefficients β_{00k} and β_{10k} are the intercept and slope and, as indicated by their subscript k, they are allowed to vary over the schools. The school level model is given by

$$\begin{aligned} \beta_{00k} &= \gamma_{000} + v_{00k} \\ \beta_{10k} &= \gamma_{100} + v_{10k}. \end{aligned} \quad (2.28)$$

Predictor variables at the class and school level may be added to models (2.27) and (2.28) to explain between-class and between-school variability of the intercept and slope.

The decomposition of the total variance in variance components is based on the empty model:

$$y_{ijk} = \gamma_{000} + v_{00k} + u_{0jk} + e_{ijk}, \quad (2.29)$$

where the random effects are independently and normally distributed with zero mean and variances σ^2_{v0}, σ^2_{u0}, and σ^2_e. The total variance of a pupil's outcome is equal to $var(y_{ijk}) = \sigma^2_{v0} + \sigma^2_{u0} + \sigma^2_e$. The covariance of two outcomes in the same class is equal to $cov(y_{ijk}, y_{i'jk}) = \sigma^2_{v0} + \sigma^2_{u0}$ and the covariance of two outcomes from different classes in the same school is $cov(y_{ijk}, y_{i'j'k}) = \sigma^2_{v0}$.

The intraclass correlation coefficients at the class and school level measure the proportion variance at these levels and are calculated as

$$\rho_{class} = \frac{\sigma^2_{u0}}{\sigma^2_{v0} + \sigma^2_{u0} + \sigma^2_e}, \quad (2.30)$$

and

$$\rho_{school} = \frac{\sigma^2_{v0}}{\sigma^2_{v0} + \sigma^2_{u0} + \sigma^2_e}, \quad (2.31)$$

see for instance Siddiqui, Hedeker, Flay, and Hu (1996). A second method to calculate the intraclass correlation coefficients exists. This method defines the intraclass correlation coefficients at the class level as the expected correlation of two randomly chosen subjects in the same class and the intraclass correlation coefficient at the school level as the expected correlation of two randomly chosen subjects in two different classes in the same school. These are calculated as:

$$\rho^*_{class} = \frac{cov(y_{ijk}, y_{i'jk})}{var(y_{ijk})} = \frac{\sigma^2_{v0} + \sigma^2_{u0}}{\sigma^2_{v0} + \sigma^2_{u0} + \sigma^2_e}, \quad (2.32)$$

and

$$\rho^*_{school} = \frac{cov(y_{ijk}, y_{i'j'k})}{var(y_{ijk})} = \frac{\sigma^2_{v0}}{\sigma^2_{v0} + \sigma^2_{u0} + \sigma^2_e}. \quad (2.33)$$

The latter pair of intraclass correlation coefficients uses an asterisk superscript to distinguish it from the first pair. It can be easily verified that $\rho^*_{class} = \rho_{class} + \rho_{school}$ and $\rho^*_{school} = \rho_{school}$.

2.6 Software for multilevel analysis

Fitting multilevel models can be done by using special purpose software or procedures in general purpose statistical programs. Among the first category, MLwiN is the most extensive program (Rasbash, Steele, Browne, & Goldstein, 2012). It uses ML and REML estimation or Bayesian methods and, in addition, it allows for data manipulation and has a graphical device to display graphs to check model assumptions. The graphical user interface makes it possible to build models interactively and knowledge of the underlying commands is not required. This interface displays the multilevel models in the combined notation as in Equations (2.6) and (2.3), so the user must write the separate equations for the separate levels and combine these in a single equation model before entering the model in MLwiN.

The program handles data structures up to five levels. It can fit models with normal and categorical outcome data, uni- and multivariate outcomes, and can be used for studies that have data that are not purely hierarchical. Such data occur when subjects change groups during the course of the study (multiple membership models), or where the grouping factor is crossed with another grouping factor (cross classified models). An example of the latter data structure is a study where pupils are nested in schools as well as in neighborhoods.

HLM (Raudenbush, Bryk, Cheong, Congdon, & du Toit, 2011) has roughly the same features as MLwiN but handles only two- and three-level data structures. It also uses a graphical user interface to interactively build the multilevel model. It builds separate models for each level, such as in Equations (2.1), (2.2) and (2.5) and it is a very useful program for beginners in multilevel modeling.

The Supermix program by Hedeker, Gibbons, du Toit, and Chen (2008) is a compilation of five Fortran programs; each one handles a specific outcome variable type: continuous, ordered, nominal, count or binary. The program can deal with two or three levels of nesting. The user interface and graphical device are less extensive than those for MLwiN and HLM. The strong feature of this program is that it uses ML estimation for all outcome variable types including non-continuous outcomes for which it uses numerical integration of the likelihood function.

Current versions of all three programs allow input of data from a variety of general purpose software packages such as SPSS and STATA.

Procedures to fit multilevel models have been incorporated in many general purpose statistical packages. The procedures MIXED and GENLINMIXED in SPSS fit linear and generalized linear multilevel models for continuous and categorical outcomes. A short introduction to using SPSS for linear multilevel models can be found in Chapter 20 of Field (2013) and Chapter 8 of Landau and Everitt (2004).

The library `nlme` in S-Plus and the corresponding package with the same name in R provide functions to fit linear and non-linear multilevel models. It allows for models with any number of levels and has extensive options to graphically display data. The drawback is that it uses syntax to specify statistical models which makes it less suitable for those unfamiliar with S-Plus and R and beginners in multilevel modeling. More information on multilevel models in S-Plus can be found in the extensive book by Pinheiro and Bates (2004). Other R packages are `nlmer` and `hglm`.

Mplus is another general purpose program that uses syntax to build statistical models (B. O. Muthén & Muthén, 2012). It is a program that was originally written for structural equation modeling but the newest versions can also fit multilevel models. It is limited to three levels of nesting for cross-sectional data and four levels for longitudinal data. It also allows multilevel data structures in path models, factor analysis models and latent class models.

The SAS program has various procedures to fit linear, generalized linear and nonlinear multilevel models: GLIMMIX, HPMIXED, MIXED and NLMIXED. These procedures can handle flexible covariance structures for random effects, which makes them very useful for repeated measures analysis. An introduction to the MIXED procedure is given by Singer (1998) but a more extensive overview of SAS for mixed models is also available (Littell, Milliken, Stroup, Wolfinger, & Schabenberger, 2006).

Stata has several commands to fit linear and generalized linear multilevel models for a wide range of outcome variable types. The procedure GLLAMM (generalized linear latent and mixed models) is available for a general class of models, including multilevel factor and item response models, multilevel structural equation models and latent class models (Rabe-Hesketh & Skrondal, 2012a, 2012b).

Latent GOLD is a program for latent class analysis, where several unobservable groups are identified. The advanced option to this program allows for two- and three-level multilevel data structures (Vermunt & Magidson, 2005).

A more extensive overview of software for linear multilevel models is given by West and Galecki (2011). For binary and ordinal outcomes, see Li, Lingsma, Steyerberg, and Lesaffre (2011) and Kim, Choi, and Emery (2013).

3

Concepts of statistical power analysis

The sample size calculations presented in this book rely on the framework of statistical power analysis. Power analysis is closely related to the statistical tool of null hypothesis testing. A review of null hypothesis testing and power analysis is given in the following section, which is followed by Sections 3.2 and 3.3 in which the different types of power analysis and the timing of power analysis are discussed. In the ideal case, simple mathematical expressions for power analysis can be derived. Unfortunately, this is not always possible and one has to rely on other methods such as simulation studies, as discussed in Section 3.4. The different factors that influence a power analysis are topics of Section 3.5. Section 3.6 provides an overview of the steps that should be taken in a power analysis. One of these steps is the choice of an appropriate research design. Some general considerations with respect to the design of experiments are given in Section 3.7. Sections 3.8 and 3.9 focus on two alternative methods for sample size calculation: precision analysis and accuracy of parameter estimates.

3.1 Background of power analysis

3.1.1 Hypotheses testing

The aim of an experimenter is to investigate whether a new treatment for some particular condition or disease performs better than a standard treatment or no treatment at all. Consider, for instance, a psychologist who wishes to study the effect of a new program that supports child rearing abilities of mothers who have sought shelter because of domestic violence. This new program is compared to a control group consisting of mothers who do not receive services through the new program. In general, a control condition consists of an existing or standard treatment, no treatment or a placebo. In the remainder of this chapter the new program condition is referred to as the experimental condition, although terms such as test condition and intervention condition are also commonly used.

Data on many outcome variables are typically collected in an intervention study; in this example the children's problem behavior is used as the primary

	decision based on statistical test	
unknown truth	no treatment effect	treatment effect exists
no treatment effect	correct decision $p = 1 - \alpha$	type I error $p = \alpha$
treatment effect exists	type II error $p = \beta$	correct decision $p = 1 - \beta =$ **power**

Figure 3.1: The outcomes of a statistical test and their probabilities.

outcome. The null hypothesis states that there is no difference between the two conditions with respect to the mean degree of problem behavior. The alternative hypothesis states that the mean degree of problem behavior in the control group is higher than that in the experimental group, which corresponds to a non-zero treatment effect. In mathematical notation the two hypotheses are formulated as

$$H_0 : \mu_C = \mu_E \text{ and } H_a : \mu_C > \mu_E. \tag{3.1}$$

Here, μ_C and μ_E are the mean scores on problem behavior in the control and experimental conditions respectively; higher scores indicate higher degrees of problem behavior. It should be noted that the outcomes are measured on a continuous scale, for instance by using sum scores of items in questionnaires. Had dichotomous outcomes been observed, proportions instead of means would have been compared. Also note that the hypotheses do not include the situation $\mu_C < \mu_E$. This is a matter of choice and implies that no statistical inference can be made about the experimental condition increasing problem behavior. If that would be also of interest, then it would make sense to phrase the alternative hypothesis as $H_a : \mu_C \neq \mu_E$. This is a case of two-sided testing to be discussed later.

Two types of correct decisions and two types of errors can be made when testing hypotheses, as is shown in Figure 3.1. A type I error occurs when the null hypothesis is incorrectly rejected. In our example this would mean that the psychologist concludes that the mean degree of problem behavior is lower in the experimental group while such a difference does not exist in the population. Such a result is called a false or fake positive result. The type I error rate is the probability of conducting such an error and is denoted α. The type I error rate is often referred to as significance level. Its complement $1 - \alpha$ is the probability of correctly not rejecting the null hypothesis. The other type of error that can be made while testing an hypothesis is referred to as type II error, and such an error is made when the null hypothesis is incorrectly not rejected. In such a case, our psychologist would conclude there is no difference in mean scores in the two conditions, while in fact the mean score is larger in the control condition than in the experimental condition. Therefore, the psychologist's conclusion is a false negative result.

The type II error rate is denoted β and its complement $1-\beta$ is the probability of correctly rejecting a false null hypothesis. This probability is called the statistical power of a test. At a fixed sample size, the type I and II error rates are inversely related: increasing the type I error rate results in a decreasing type II error rate and vice versa. In practice, small values for both are desired. The type I error rate is controlled up front by selecting the appropriate quantile of the distribution of the test statistic under the null hypothesis, as explained below. Given a value of the type I error rate, the type II error rate is controlled by selecting the appropriate sample size.

3.1.2 Power calculations for continuous outcomes

In order to test whether the null hypothesis is true, test subjects are randomly drawn from a population of mothers who have sought refugee at a women's shelter and are randomly assigned to treatment conditions. After the intervention has been implemented, their children's degrees of problem behavior are measured and $y_{C,1}, ..., y_{C,i}, ..., y_{C,n_C}$ and $y_{E,1}, ..., y_{E,i}, ..., y_{E,n_E}$ are the n_C and n_E responses in the control and experimental groups. These responses are assumed to be independently and normally distributed. For simplicity, the standard deviations σ_C in the control group and σ_E in the experimental group are assumed known. The z-test for two independent samples is used to test whether the mean difference $\mu_C - \mu_E$ differs from zero. The test statistic is given by

$$\frac{\overline{y}_C - \overline{y}_E}{\sqrt{\frac{\sigma_C^2}{n_C} + \frac{\sigma_E^2}{n_E}}}, \tag{3.2}$$

where $\overline{y}_C = \Sigma_i y_{Ci}/n_C$ and $\overline{y}_E = \Sigma_i y_{Ei}/n_E$ are the average scores in both treatment groups. The numerator is the point estimator of the treatment effect and the denominator its standard error.

The null hypothesis is rejected when the test statistic exceeds a critical value, as depicted in the upper half of Figure 3.2. The probability of incorrectly rejecting the null hypothesis is fixed at the significance level α, which is equal to the proportion of the distribution of the test statistic under the null hypothesis that is above the critical value. This probability is indicated by the grey area. Under the null hypothesis of no treatment effect the test statistic (3.2) has a standard normal distribution z. Thus the critical value is equal to $z_{1-\alpha}$, which is the $100(1-\alpha)$th percentile of the standard normal distribution.

Under the alternative hypothesis the test statistic (3.2) follows a normal distribution with mean value $(\mu_C - \mu_E)/\sqrt{\sigma_C^2/n_C + \sigma_E^2/n_E}$, where the difference $\mu_C - \mu_E$ is the size of the treatment effect. It is often referred to as the unstandardized effect size and reflects the difference in mean scores of the two conditions. The distribution of the test statistic under the alternative hypothesis is shifted to the right as compared to the distribution under the null hypothesis; see the lower half of Figure 3.2. The statistical power $1 - \beta$ is the probability that the test statistic is above the critical value $z_{1-\alpha}$ when

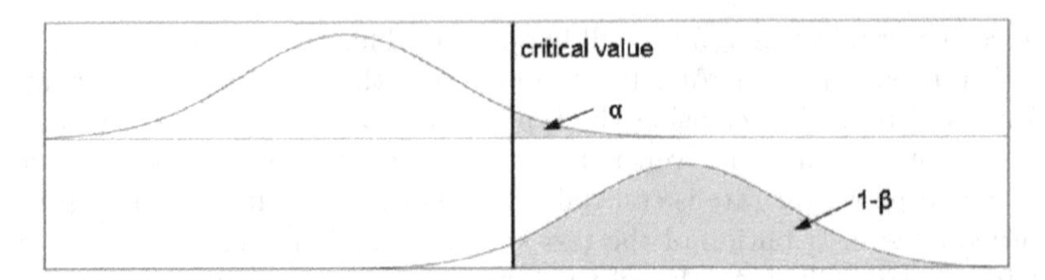

Figure 3.2: Distribution of the test statistic under the null (top) and alternative (bottom) hypotheses and the probability of a type I error α and the statistical power $1 - \beta$.

the alternative hypothesis is true. It is equal to

$$P\left(z + \frac{(\mu_C - \mu_E)}{\sqrt{\frac{\sigma_C^2}{n_C} + \frac{\sigma_E^2}{n_E}}} > z_{1-\alpha}\right) = P\left(z > z_{1-\alpha} - \frac{(\mu_C - \mu_E)}{\sqrt{\frac{\sigma^2}{n_C} + \frac{\sigma^2}{n_E}}}\right), \tag{3.3}$$

where the random variable z has the standard normal distribution. The desired power level $1 - \beta$ is achieved when

$$z_{1-\alpha} - \frac{(\mu_C - \mu_E)}{\sqrt{\frac{\sigma_C^2}{n_C} + \frac{\sigma_E^2}{n_E}}} = -z_{1-\beta}. \tag{3.4}$$

In case of a two-sided alternative hypothesis $H_a : \mu_C \neq \mu_E$ the term $z_{1-\alpha}$ is replaced by $z_{1-\alpha/2}$. From Equation (3.4) explicit formulae for the required sample sizes n_C and n_E can be derived. For instance, if the variance is constant across groups ($\sigma_C^2 = \sigma_E^2 = \sigma^2$) and group sizes are equal ($n_C = n_E = n$), then

$$n = \frac{2\sigma^2(z_{1-\alpha} + z_{1-\beta})^2}{(\mu_C - \mu_E)^2}. \tag{3.5}$$

The group size calculated from this equation is often a non-integer number and needs to be rounded upward. The achieved power will be slightly higher than the desired power level. In case the variance varies over groups, Cohen (1988) suggests using the mean variance $\sigma^2 = (\sigma_C^2 + \sigma_E^2)/2$; other approaches are to replace σ^2 in Equation (3.5) by the variance in either the control or experimental condition. An even better procedure would be to impute the values of the variances σ_C^2 and σ_E^2 in Equation (3.4) and to calculate the sample sizes n_E and n_C for which the desired power level is achieved.

As the sample sizes n_E and n_C are the two unknowns in Equation (3.5) one should pose an additional restriction on these sample sizes. For instance, one could fix the sample size in one treatment arm and calculate the other, or one could fix the ratio n_E/n_C of the sample sizes. For instance, the ratio is equal to 2 if one wishes to include twice as many subjects in the experimental condition as in the control condition.

In practice, the standard deviations of the responses are often unknown and replaced by estimates from the collected data. The difference in means is then tested using an independent samples t test which follows a central t distribution with $df = n_C + n_E - 2$ degrees of freedom under the null hypothesis. The critical value is then equal to $t_{n_C+n_E-2,1-\alpha}$, which is the $100(1-\alpha)$th percentile of the t distribution with n_C+n_E-2 degrees of freedom. Under the alternative hypothesis this test statistic follows a non-central t distribution with non-centrality parameter $\lambda = (\mu_C - \mu_E)/\sqrt{\sigma_C^2/n_C + \sigma_E^2/n_E}$ and $n_C + n_E - 2$ degrees of freedom. The power is equal to the probability that this test statistic exceeds the critical value:

$$P\left(t_{n_C+n_E-2}(\lambda) > t_{n_C+n_E-2,1-\alpha}\right), \tag{3.6}$$

where $t_{df}(\lambda)$ is a random variable that follows a non-central t distribution with df degrees of freedom and non-centrality parameter λ. This formula replaces Formula (3.3). The larger the sample sizes, the larger the non-centrality parameter, and hence the larger the power of the test.

An explicit formula for the desired sample size cannot be given since the sample sizes n_C and n_E appear in both the non-centrality parameter and the degrees of freedom. However, the power for a range of sample sizes can be easily calculated using a spreadsheet and the minimal sample size for which a desired power level is achieved can be selected from this range. Cousineau (2007) shows how to perform such calculations in SPSS or Mathematica.

Equation (3.6) holds for a test with a one-sided alternative hypothesis $H_a : \mu_C > \mu_E$. For an alternative hypothesis $H_a : \mu_C < \mu_E$ the power is calculated as

$$P\left(t_{n_C+n_E-2}(\lambda) < t_{n_C+n_E-2,\alpha}\right), \tag{3.7}$$

and for a two-sided alternative $H_a : \mu_C \neq \mu_E$ the relation

$$P\left(t_{n_C+n_E-2}(\lambda) < t_{n_C+n_E-2,\alpha/2}\right) + P\left(t_{n_C+n_E-2}(\lambda) > t_{n_C+n_E-2,1-\alpha/2}\right) \tag{3.8}$$

is used. The power is now the sum of two probabilities. In cases λ is positive the first probability is relatively small to the second probability, and vice versa when λ is negative. When λ is large, one of the probabilities will be ignorable compared to the other and for convenience it is often ignored in the calculation of the power level. Alternatively one can calculate the power level on basis of an F test in case an two-sided alternative is used

$$P\left(F_{1;n_C+n_E-2}(\lambda^2) > F_{1;n_C+n_E-2,1-\alpha/2}\right). \tag{3.9}$$

The degrees of freedom are equal to 1 and $n_C + n_E$. Note that the square of the non-centrality parameter λ is used.

In case the standard deviations are unknown, prior estimates can be obtained from the literature or a pilot study to calculate the required sample size of a future trial. It should be noted that these estimates are associated with imprecision as they vary from sample to sample. Vickers (2003) shows that the standard deviations estimated in a trial were very often larger than the prior estimates used in the sample size calculations, which resulted in many underpowered studies. Underestimation of the standard deviations may be caused by a small sample size in a pilot study or by using selective data. For instance, estimating the standard deviations in a multisite trial on basis of data from just one site may result in underestimation as subjects within one site are often more similar than subjects from different sites.

Vickers (2003) provides correction factors for estimates obtained from pilot data, based on formulae from R. H. Browne (1995). Atenafu, Hamid, Stephens, To, and Beyene (2009) caution against using p-values of published research to determine sample size for future research. Trials with small sample sizes may have very low p-values, which are often caused by chance. Atenafu et al. (2009) show that studies with very small p-values may not have adequate power, and using the same sample size in future studies may result in not detecting an effect of treatment.

G. F. Borm, Bloem, Munneke, and Teerenstra (2010) focus on sample size calculations based on insufficient information from previous trials in the literature. For instance, the power calculation for an outcome variable that measures change from baseline requires the standard deviation of the difference, while often only the standard deviations of the baseline and follow up measurements are reported. The method proposed by G. F. Borm et al. (2010) to calculate power is solely based on the p-value or confidence interval of the reference study.

The problem of unknown standard deviations is less important if the unstandardized effect size $\mu_C - \mu_E$ is replaced by the standardized effect size δ, which is calculated as $\delta = (\mu_C - \mu_E)/\sigma_p$, where σ_p is the pooled standard deviation. The non-centrality parameter of the t distribution is then equal to $\delta/\sqrt{1/n_C + 1/n_E}$, which does not depend on σ. In other words, the power does not depend on σ once a standardized effect size is used.

The standardized effect size represents the difference in mean scores between the two programs expressed in standard deviation units, and is scale free. When using the standardized effect size, one thinks of the effect not in terms of the underlying scale, but as a multiple or fraction of the (unknown) standard deviation. The advantage is that one does not have to decide what the effect and the standard deviation on the underlying scale would have to be. The other side of the coin is that the relevance of a standardized effect of a certain magnitude still has to be discussed (i.e., after the experiment has been conducted). A zero effect size corresponds to the null hypothesis of no treatment effect.

Cohen (1988) makes a distinction between small, medium and large effect sizes. Small effects are often encountered when studying outcomes that are not under good experimental or measurement control or both. In such cases, the "signal" $\mu_C - \mu_E$ is small in relation to the "noise" σ as produced by uncontrollable extraneous variables, which results in a small effect. A medium effect size is sufficiently large to be visible by the naked eye of a careful and experienced researcher. The difference between both treatment conditions with respect to their mean outcomes is obvious when the effect size is large. In our example where two mean values of a continuous outcome variable are compared, $\delta = 0.2, 0.5$, and 0.8 are regarded as small, medium and large. A further introduction to effect sizes can be found in Cohen (1988), Grissom and Kim (2005), and Lipsey and Wilson (1993).

3.1.3 Power calculations for dichotomous outcomes

Many outcomes in cluster randomized trials are measured on a binary scale and the outcomes variables are dichotomous rather that continuous. Such outcomes measure whether a particular event has occurred. In clinical trials, one may investigate whether a patient has recovered or not, in substance abuse trials one may measure whether one uses alcoholics, cigarettes or another type of substance and in trials on the effects of employment and training programs one may investigate whether someone who receives social benefits has found a paid job. The corresponding null hypothesis is $H_0 : p_C = p_E$, where p_C and p_E are the probabilities in the control and experimental conditions, respectively. Various effect sizes for trials with dichotomous outcomes exist, of which the risk difference $p_C - p_E$, risk ratio or relative risk p_C/p_E and odds ratio $(p_E(1-p_C))/(p_C(1-p_E))$ are the most commonly used. For small probabilities p_E and p_C, the relative risk and odds ratio are almost equal. The risk difference and odds ratio are used in the sample size calculations that follow.

3.1.3.1 Risk difference

Various test statistics are available to test the null hypothesis of no risk difference $H_0 : p_C - p_E = 0$, such as the binomial test, Fisher's exact test, the likelihood ratio test and asymptotic test statistics that rely on large sample sizes. Asymptotic test statistics are used to perform a power analysis in this section and all chapters that follow as the other test statistics are not well studied for trials with nested data. An introduction to these test statistics for non-nested data can be found in Chapters 5 and 6 of Chow et al. (2008).

The probabilities p_C and p_E are estimated by the proportions $\hat{p}_C$ and $\hat{p}_E$. An estimate of the risk difference $p_C - p_E$ is $\hat{p}_C - \hat{p}_E$, which has asymptotic standard error

$$s.e.(\hat{p}_C - \hat{p}_E) = \sqrt{\frac{p_C(1-p_C)}{n_C} + \frac{p_E(1-p_E)}{n_E}}, \tag{3.10}$$

with n_C and n_E the sample sizes in the control and experimental condition. This standard error can be estimated by replacing the probabilities p_C and p_E by their estimates $\hat{p}_C$ and $\hat{p}_E$. Given large sample sizes, the test statistic

$$\frac{(\hat{p}_C - \hat{p}_E)}{\hat{s.e.}(\hat{p}_C - \hat{p}_E)} \tag{3.11}$$

follows a standard normal distribution under the null hypothesis. As for continuous outcomes, the power to detect a positive difference $p_C - p_E > 0$ is the probability that this test statistic exceeds the critical value $z_{1-\alpha}$:

$$P\left(\frac{p_C - p_E}{\sqrt{\frac{p_C(1-p_C)}{n_C} + \frac{p_E(1-p_E)}{n_E}}} > z_{1-\alpha}\right). \tag{3.12}$$

The desired power level is then achieved when

$$z_{1-\alpha} - \frac{p_C - p_E}{\sqrt{\frac{p_C(1-p_C)}{n_C} + \frac{p_E(1-p_E)}{n_E}}} = -z_{1-\beta}. \tag{3.13}$$

For equal group sizes $n_C = n_E = n$ the required number of subjects per group is equal to

$$n = \frac{(p_C(1-p_C) + p_E(1-p_E))(z_{1-\alpha} + z_{1-\beta})^2}{(p_C - p_E)^2}. \tag{3.14}$$

For unequal group sizes one may either fix the group size in one treatment arm and calculate the other or fix the ratio n_E/n_C of the group sizes. The sample sizes can then be calculated from Equation (3.13). This equation assumes the true values p_C and p_E to be known. This is generally not the case in the design phase of a trial and they should be replaced by values that are based on the researchers' expectations or estimates from similar studies in the past.

3.1.3.2 Odds ratio

Power calculations for dichotomous outcomes can also be based on the odds ratio rather than on the risk difference. The odds for group i are defined as $p_i/(1-p_i)$, hence the odds ratio between the experimental and control condition is given by

$$OR = \frac{p_E(1-p_C)}{p_C(1-p_E)}, \tag{3.15}$$

where, once again, p_C and p_E are the probabilities in the control and experimental condition, respectively. The odds ratio is always larger than 0 and usually ranges between 0 and 4 (Chow et al., 2008). An odds ratio equal to 1 implies there is no difference between the two treatments with respect to the dichotomous outcome. An odds ratio larger than 1 implies the subjects in the experimental condition are more likely to have the outcome of interest than those in the control condition, whereas an odds ratio between 0 and 1 implies the opposite.

The natural logarithm of the estimated odds ratio has asymptotic standard error

$$s.e.(log(\widehat{OR})) = \sqrt{\frac{1}{p_C(1-p_C)n_C} + \frac{1}{p_E(1-p_E)n_E}}. \tag{3.16}$$

The null hypothesis of no treatment effect is tested with the test statistic $log(\widehat{OR})/s.e.(log(\widehat{OR}))$, which asymptotically follows the standard normal distribution under the null hypothesis. The desired power level is achieved when

$$z_{1-\alpha} - \frac{log(OR)}{\sqrt{\frac{1}{p_C(1-p_C)n_C} + \frac{1}{p_E(1-p_E)n_E}}} = -z_{1-\beta}. \tag{3.17}$$

For equal numbers of subjects in both treatment arms ($n_C = n_E = n$), the size per treatment arm is given by

$$n = \frac{(\frac{1}{p_C(1-p_C)} + \frac{1}{p_E(1-p_E)})(z_{1-\alpha} + z_{1-\beta})^2}{(OR)^2}. \tag{3.18}$$

To be able to calculate the required sample size, estimates for the probabilities p_C and p_E should be provided based on expert knowledge or findings from the literature.

The approach based on the odds ratio is directly related to the logistic regression model. The model for subject i is given by

$$p_i = \frac{1}{1 + e^{-(\gamma_0 + \gamma_1 x_i)}}, \tag{3.19}$$

where x_i indicates treatment condition and the treatment effect γ_1 is the natural logarithm of the odds ratio. Given coding $x_i = -0.5$ for the experimental group and $x_i = 0.5$ for the control group, a positive γ_1 implies that the probability of an outcome $y_i = 1$ is higher in the experimental group than in the control group, and a negative γ_1 implies the opposite.

The approaches based on the risk difference and odds ratio give rather similar values of the required sample sizes. The advantage of the latter approach is that it can be extended by including background variables and risk factors as covariates in the logistic regression model.

3.2 Types of power analysis

Equation (3.4) holds for testing means in trials with independent samples and continuous outcomes. Its general equivalent for any type of trial is given by

$$\frac{\text{effect size}}{\text{standard error}} \approx z_{1-\alpha} + z_{1-\beta}, \tag{3.20}$$

for a one-sided alternative hypothesis. For a two-sided alternative, $z_{1-\alpha}$ is replaced by $z_{1-\alpha/2}$, where the generally negligible probability of a rejection at the wrong side of the critical region is ignored (see Equation (3.8) and the succeeding text). This approximation is based on the fact that the sampling distribution of maximum likelihood estimates is asymptotically normal and is therefore valid when the sample size is sufficiently large. A relation between four parameters is observed: the significance level α, the power $1-\beta$, the effect size and its related standard error. The standard error is inversely related to the sample size: the larger the sample size, the lower the standard error. Once three of these four parameters are specified, the fourth can be calculated on basis of Equation (3.20). Thus, four types of power analyses can be distinguished (Cohen, 1988).

Compute sample size as a function of power, significance level and effect size. This type of power analysis enables the researcher to calculate the required sample size before the experiment is conducted. It is recommended when the number of subjects that can be recruited is not limited by financial or practical constraints.

Compute power as a function of sample size, significance level and effect size. This type of power analysis can be used to compute the probability which a given statistical test has to detect a prespecified effect size at a given sample size.

Compute significance level as a function of power, sample size and effect size. This type of power analysis is reasonable when the control of the significance level is less important than the control of the power and is referred to as criterion analysis. It is very uncommon in practice since researchers are often not willing to use large values of the significance level.

Compute effect size as a function of power, significance level and sample size. This type of power analysis is referred to as sensitivity analysis. It enables the researcher to investigate whether the effect size that can be detected with a proposed study at a given sample size is relevant, or it can be used to calculate the effect size that a published study was able to detect with a certain power level.

Although four types of power analyses exist, this book primarily focuses on the first and second and studies the relationship between power and sample size at a given significance level and effect size. The other two types of power analyses can easily be performed once the relation between the standard error of the estimated treatment effect and sample size is known.

3.3 Timing of power analysis

Besides the four different types of power analyses (sample size calculation, power calculation, criterion analysis, sensitivity analysis) discussed in Section 3.2, the timing of the power analysis can be distinguished: pre-study or post-study. Strictly speaking, during-study could also be an option, for example a sample size re-estimation in the course of a trial (Lake, Kammann, Klar, & Betensky, 2002), but this will not be discussed here.

Post-study power analyses are performed after the data has been analyzed and hence when the effect and its variance in the study have been observed. Pre-study power analyses are performed before analyzing the data, hence without knowing the actual results of the study. A variety of terms are used, but a priori power analyses are synonyms for pre-study power analyses and observed power and retrospective power analyses are synonyms for post-study power analyses.

Some ambiguity is in the use of post hoc and a posteriori power analyses. Sometimes this terminology is used in the planning stage, when power at varying sample sizes (or varying effect sizes) is investigated. Then this analysis is post hoc or a posteriori (Latin for afterwards) only in the sense that first a sample size (or effect size) is given and then the power is calculated from that sample size or effect size. More logically (and more often), however, post hoc and a posteriori mean "after the study data has been analyzed" and then these refer to post-study power analyses.

Distinguishing the timing of the power analysis is important for the following reason. The use of and rationale for a priori power analyses has been accepted almost universally (Section 1.4). This is not the case for post hoc (in the sense of post-study) power analyses. In its simplest form, a post hoc power analysis is performed when the observed effect is not statistically significant (but of an interesting magnitude) and proceeds as follows.

The power is calculated using both the effect and its variance as observed in the study and then back-interpreted to the conducted study in the following way. If the post hoc power for the observed effect is low, then the non statistical significance is (implicitly) considered questionable, that is, the observed effect is considered to indicate a real effect and the lack of statistical significance is considered due to the low post hoc power of the study.

The rationale for this practice is controversial. Proponents such as Onwuegbuzie and Leech (2004, P. 221) argue that post hoc power should be calculated to avoid type II errors (the test is not statistically significant, while the effect is real) or so-called type A errors (the effect is large, but not statistically significant). Opponents, for example, Thomas (1997), have argued that type A errors can also be addressed using the confidence interval of the observed effect which would be expected to extend to large values where the true effect is large.

Hoenig and Heisey (2001) show that a post hoc power analysis based on the observed effect and observed variances is a way of restating the p-value (or of restating the observed effect with its confidence interval) because the power and p-value then have a one-to-one correspondence. As an example they elaborate for a one-sided z-test, that the one-to-one correspondence between post hoc power and p-value is as follows: the post hoc power is less than 0.50 if $p > 0.05$ and vice versa.

Therefore, the above type of post hoc power analysis does not add information to what is already known and as a consequence, it cannot address the problem of type II errors. The information that is in a completed study is already told by the observed effect with its confidence interval. Most importantly, the observed effect and variance (with their confidence intervals) are useful to plan future studies. Other illustrations of controversial use or misuse of post hoc power are discussed in Hoenig and Heisey (2001) and Thomas (1997).

3.4 Methods for power analysis

Two methods for power calculation exist: an analytical approach and a power analysis by means of a simulation study. The analytical approach aims to derive a mathematical relationship of the standard error of the treatment effect as a function of the sample size, possibly using large sample approximations. Once such a relationship is available, the standard error can be calculated for a given sample size and imputed into the left side of Equation (3.20) to calculate the corresponding power. For simple designs, the standard error can easily be obtained on basis of simple mathematical derivations. Consider a trial with two treatment conditions and independent and continuous outcomes. The average scores $\overline{y}_C$ and $\overline{y}_E$ are unbiased estimates for the population means and their difference $\overline{y}_C - \overline{y}_E$ is an unbiased estimate for the effect size. The average scores have variances σ_C^2/n_C and σ_E^2/n_E and, as the outcomes are independent, the variance of the effect size is equal to $\sigma_C^2/n_C + \sigma_E^2/n_E$. The standard error is calculated by taking its square root.

For more complicated designs a mathematical expression for the standard error of the effect size can be calculated by using matrix-vector algebra. Examples are trials with more than two treatments or factorial designs with or without interactions between the factors. In addition, matrix-vector algebra is also useful when covariates such as baseline scores and risk factors are expected to have an effect on the outcomes variable, or when the effect of treatment condition is moderated by one or more covariates. A regression model is formulated to express the relation between the outcome variable and treatment condition(s), covariates, and their interactions.

This model is expressed in matrix vector notation and matrix algebra is used to derive a mathematical expression of the covariance matrix of the estimator of the regression coefficients. The standard error of the treatment effect estimator is part of this matrix and can be used to calculate the required sample size to achieve a desired power level. Matrix-vector algebra is used to derive mathematical expressions for the standard error of the treatment effect estimator for most of the designs discussed in the following chapters.

A simulation study is another method to perform a power analysis and relies on computer-intensive methods. It is also referred to as a Monte Carlo study. This procedure is particularly useful if the regression model that relates outcome to response is non-linear, such as a logistic regression model for dichotomous or categorical responses, or when the assumption of independent outcomes is violated. For such models, the derivation of a simple explicit formula for the variance of the treatment effect estimator as a function of sample size is not always an easy task.

The procedure for a simulation study is as follows. Given a regression model, values of all model parameters, and a sample size, a large number of data sets is simulated. Nowadays fast computers allow many data sets to be generated in a short time and the number of data sets is typically set to 1000, 5000, or even 10,000. For each simulated data set, the model parameters and their standard errors are estimated and the observed power is calculated as the proportion of data sets for which the null hypothesis of no treatment effect is rejected. This procedure is repeated for a range of sample sizes to gain insight in the relation between sample size and power.

L. K. Muthén and Muthén (2002) propose the following strategy for deciding on sample size based on three criteria concerning the parameter and standard error bias and coverage of the 95% confidence interval. The first criterion states that the parameter and standard error bias for each parameter θ in the model should be less than 10%. The percentage parameter bias is calculated as $100\%(\bar{\hat{\theta}} - \theta)/\theta$ where θ is the true value of the model parameter and $\bar{\hat{\theta}}$ is its mean estimate over all generated data sets. The percentage standard error bias is calculated as $100\%(\overline{se}(\hat{\theta}) - sd(\hat{\theta}))/sd(\hat{\theta})$ where $\overline{se}(\hat{\theta})$ is the mean of the standard errors of the estimates $\hat{\theta}$ and $sd(\hat{\theta})$ is the standard deviation of all estimates $\hat{\theta}$. The second criterion states that the standard error bias for the parameter of interest (i.e., the treatment effect in a trial) is less than 5%. The third criterion states that the coverages of the 95% confidence intervals are between 0.91 and 0.98. Once these criteria are satisfied, the sample size is chosen such that the observed power is close to the desired value.

A simulation study can be used to perform power analyses for many types of multivariate statistical models, including complex growth models and path models where one or more of the variables are latent and measured by some indicator variables. To be able to perform the power analysis, one must specify the underlying statistical model and provide the population values of all model parameters, including factor loadings, (co)variances of exogenous variables and error variances. In practice, this information is rarely known and often not available from the literature. The same statistical model and measurement instruments are seldom used. In such cases, one must provide the most plausible statistical model and plausible values of its parameters. It should be clear this is not an easy task that can be performed just a few days prior to deadlines of grant proposals. In fact, a power calculation can be a study in its own right and a sufficient amount of time should be devoted to it.

It should finally be mentioned that a power analysis is just one of the applications of simulation studies. Simulation studies are often performed to study the performance of estimation methods or the robustness of statistical models against violations of model assumptions. General guidelines with respect to simulation studies can be found in Arnold, Hogan, Colford, and Hubbard (2011), Boomsma (2013), Burton, Altman, Royston, and Holder (2006), Landau and Stahl (2013) and Paxton, Curran, Bollen, Kirby, and Chen (2001).

3.5 Robustness of power and sample size calculations

Different factors influence the power (and sample size) calculation. The power calculation is based on the distribution of the test statistic under the alternative (and null) hypothesis. These distributions are often derived theoretically or via simulations based on assumptions on how the mean (or whatever statistic is being estimated) is related to the treatment conditions and assumptions on the error distribution. For example, in linear multilevel models, the mean is linearly related to the treatment conditions (linear link function) and the error is normally distributed. In logistic multilevel models, the proportion is related to the treatment condition via the logistic link (i.e., log of the odds) and the error distribution is binomial. As a consequence, if the model (the relation between treatment conditions and the statistic to be estimated and/or the error distribution) is misspecified, it may affect power because the standard errors are not correct. Furthermore, the distributions of the test statistic may be known (or have an accessible form) only asymptotically, that is, for large sample sizes. In that case, asymptotic formulae for standard errors may not be accurate for small samples and a simulation study is needed to investigate the power for small samples.

Another influence on the power is the choice of α and the assumptions on the effect size. Also the assumptions on the standard deviation and correlations matter as these also determine the standard error of the estimator of the effect. The uncertainties in the effect size or correlations and standard deviation are of greater influence on power than the choice of α. This is clearly illustrated in formula (3.20), which can be rewritten as $z_{1-\beta} = (\text{es}/\text{se}) - z_{1-\alpha}$, where es is the effect size and se is the standard error. Thus, if the power $= 1-\beta = 0.8$ and the significance level $\alpha = 0.05$ (one-sided), then $0.84 = \text{es}/\text{se} - 1.64$ so $\text{es}/\text{se} = 2.48$. Thus, halving α (to 0.025) gives: $\text{es}/\text{se} - 1.95 = 2.48 - 1.95 = 0.53$, so power is 70.07%. Doubling α to 0.10 gives 88.58% power. In contrast, halving effect size or doubling the standard error gives 34.39% power and doubling effect size or halving standard error gives 99.95% power.

Often it is hard to have firm knowledge of the correctness of the model, and more so, of the effect size, correlations and standard deviations. In addition to that, extra assumptions may have to be made, such as large sample approximations or simulations under given assumptions, to make power and sample size calculations doable. Therefore, it makes sense to vary these assumptions over a sensible range and see how this affects the power and sample size. These considerations, in addition to feasibility considerations, may then motivate the choice of sample size.

3.6 Procedure for a priori power analysis

This section illustrates the procedure for an a priori power analysis to calculate power as a function of sample size (Steps 1 through 6) or to calculate the required sample size (Steps 1 through 7) to achieve a prespecified level of power in the planning phase of a trial. The required sample size does not only depend on the chosen significance level, the desired power level and the effect size, but also on the trial design, and measurement level of the outcome variables. Choices on these are made in the following steps.

Step 1. Decide on outcome variables and their measurement levels. A typical trial in the social and medical sciences has multiple outcome variables. A choice on the primary outcome variables and the appropriate instruments to measure them should be made in this step. Social scientists typically use questionnaires, interviews or observations to measure constructs such as antisocial behavior, depression and problem behavior. Medical scientists use equipment and laboratory tests to measure physical parameters such as hormone levels and tumor size. A related choice concerns the levels at which outcome variables are measured. For instance, a dichotomous outcome variable measures whether a subject exhibits a particular type of behavior or not, an ordinal outcome variable uses ordered categories to describe the severity of a subject's behavior and a continuous outcome variable assigns a score to a subject's behavior. Continuous outcomes may be dichotomized on basis of a cut-off point or percentile to distinguish severe from non-severe cases. The choice of the measurement level is important since it affects the choice of the type of regression model to relate outcome to treatment condition, as is explained in Step 3.

A researcher should also choose which covariates should be included in the trial and which measurement instruments and levels should be used to measure them. Covariates are variables that are expected to have an effect on the outcome variable and are included as predictor variables in the regression model. Typical covariates are the pre-test measurement on the outcome variable, gender, age and socio-economic status. The treatment groups may be expected to be balanced with respect to covariates when the sample size of a trial is large and subjects are randomly assigned to treatment conditions. In that case, the purpose of including covariates in the regression model is to explain part of the error variance and thus enhance statistical power. Random assignment of subjects to treatment conditions does not guarantee the treatment groups are comparable with respect to covariates when the sample size is small. Covariates are then included in the regression model to account for pre-treatment differences between treatment groups.

Step 2. Select an appropriate research design. Research design refers to the structure of a study. Often different research designs are applicable to a trial at hand and a choice should be made based on practical and statistical criteria. Consider a clinical trial that aims to compare the effects of two programs on some particular outcome variable. Randomization to treatment conditions can be done at the clinic level or the patient level. In the first, case all patients within the same clinic receive the same treatment condition, which may reduce administrative and logistical costs. Furthermore, health professionals should be trained in delivering only one treatment condition, which may also have a decreasing effect on costs. On the other hand, both treatment conditions are available in each clinic when randomization is done at the patient level. This design can be shown to have a higher power level to detect a treatment effect where control group contamination is absent. This assumption is reasonable when the treatments are injections or pills that only differ in their amount of active substance but it may not hold when the treatments rely on interpersonal interactions. In guideline trials, for instance, contamination may be due to the person delivering the treatment. Consider for instance a trial where physicians are educated about guidelines to reduce unhealthy lifestyles. It would be difficult for the physician to let only patients in the intervention condition benefit from the education. In such cases, the appropriate unit of randomization is the physician and not the patient.

Whichever research designs are applicable to a particular trial, the pros and cons of each of them should be weighed on the basis of ethical, practical, political, financial and statistical criteria to justify the selected research design.

Step 3. Choose a statistical model. A regression model that relates the outcome variable to treatment condition and covariates is used to estimate and test whether a difference between treatment conditions exists. The appropriate regression model is determined by the choices made in the previous two steps. For instance, a linear model should be selected when outcomes are measured on a continuous scale, whereas a generalized linear model is appropriate when outcomes are categorical. The chosen research design should also be reflected in the statistical model. For instance, complete groups of subjects are randomized to treatment conditions where a cluster randomized trial is chosen. The traditional regression model assumes uncorrelated responses and should be replaced by the multilevel model when subjects are nested within groups with the purpose to correctly model dependence of outcomes within the same group. The choice of the correct regression model is important since it not only has an effect on the estimate and test of the treatment effect in the data analysis stage, but also on the calculation of the required sample size in the design stage.

Step 4. Prespecify the effect size for the power analysis. Equation (3.20) shows that the sample size to achieve a desired power level depends on the effect size. Providing the population value for the effect size is not always an easy task. Remember that an experiment is planned to gain insight in the values of the model parameters, in particular the size of the treatment effect, while the population value of this parameter needs to be known beforehand to perform a power calculation. This introduces a vicious circle from which one can escape by replacing the population value by an educated guess obtained from historical data, a pilot study or expert knowledge. In this case, the aim is to confirm the educated guess of the population value. Alternatively, the aim may be to investigate whether a relevant effect size is present in the population studied. In this case, one needs to think what would constitute the *minimal relevant* value of the effect size in the population at hand. This requires subjective criteria from both the researchers and the subjects in the trial. For instance, the population could be obese patients and the effect could be weight loss due to stomach reduction (experimental condition) versus no stomach reduction (control condition). Perhaps one obese patient would be willing to undergo a potential harmful stomach reduction to lose 10kg, while the other patient would only be willing to do so if the loss in body weight is several dozen kilograms. The chosen value of the minimal relevant effect size should therefore be carefully justified.

Step 5. Select the desired power level and type I error rate. A choice should be made on the type I error rate or significance level α. Stringent criteria for α result in a low probability of detecting a treatment effect, especially when this effect is small and few subjects are included in the trial. On the other hand, the null hypothesis of no treatment effect is easily rejected when lenient criteria are used. Scientists often tend to use traditional values that are accepted in their field, such as $\alpha = 0.1$, 0.05, or 0.01, but they should carefully choose a value that reflects the consequences of incorrectly detecting a treatment effect, while such an effect does not exist in the population. The same reasoning applies to the choice of the power level. Although some scientists consider power moderate when it is 0.50 and high when it is at least 0.8, the choice should be matched to the study at hand and should reflect the consequences of incorrectly failing to detect an existing treatment effect. The power calculation should also include the number of primary endpoints a study has. For instance, if a study has two primary endpoints, each endpoint should have at least 90% power to preserve the overall power of the study (G. F. Borm et al., 2006).

Step 6. Calculate power as a function of sample size. Given the statistical model selected in Step 3, the effect size selected in Step 4 and the type I error rate selected in Step 5, the relation between sample size and power can be established on basis of the methods discussed in Section 3.4. If an analytical approach is possible, it may result in a formula for power given sample size.

Step 7. Select the required sample size. The sample size for which the desired power level as selected in Step 5 is achieved is determined in this final step. An analytical approach leading to a formula for power as a function of sample size can be inverted to a sample size formula. However, if a simulation study was needed, typically the power for a range of sample sizes should be investigated to determine which sample size gives the required level of power.

In general, the sample size consists of multiple components. In our example trial that aims to estimate and test the effect of a new program to support mothers in their child rearing behavior, a choice should be made on the ratio of the number of subjects in both treatment conditions. This choice can be based on criteria of a financial, ethical or statistical nature. For instance, one may choose to allocate fewer subjects to the experimental condition if this condition is considerably more expensive than the control condition, if it is expected to have adverse side effects, or if the number of professionals trained to deliver the experimental condition is limited. Statistical criteria for selecting the optimal allocation ratio are further discussed in Section 3.7.

3.6.1 An example

To illustrate the use of the seven steps described above, the example introduced in Section 3.1 is further elaborated. A psychologist wishes to evaluate the effect of an intervention that aims to reduce problems among children whose mothers have sought refuge for domestic violence. A typical instrument to measure children's problem behavior is the *Child Behavior Checklist* (CBCL) (Achenbach, 1991). It consists of 112 items related to internalizing behavior, externalizing behavior and other behaviors. Power calculations are performed for the externalizing disorder scale, which consists of 33 items that measure aggressive and delinquent behavior. The scores on these items range from 0 (not true) to 2 (very true or often true). Raw total scores are transformed to normalized T scores using procedures described in test manuals. The value 60 is used to distinguish clinical from subclinical levels of problem behavior. In the remainder, power calculations are performed for the T score by treating it as a continuous variable and for the dichotomous variable that measures whether a child exhibits clinical levels of problem behavior (Step 1).

The parallel group design is used so mothers are randomized to the experimental or control condition (Step 2). If a family consists of more than one child, the youngest child is selected as the target. The independent samples t test is used to relate the total scores to treatment condition, whereas the risk difference is used for the dichotomous version of the outcome variable (Step 3). As only one child per family is selected, the nesting of children in families can be ignored in the statistical analysis. Had more than one child been selected, the multilevel linear and logistic models should have been used.

As the true values of the mean total scores and proportions of children with clinical levels of problem behavior are unknown in the design stage, estimates are chosen from a trial by McDonald, Jouriles, and Skopp (2006). Mean T scores in the experimental and control condition are estimated to be equal to $\overline{x}_1 = 60.0$ and $\overline{x}_2 = 54.5$, so the estimated unstandardized effect size is $\mu_C - \mu_E = 5.5$. The corresponding standard deviations are equal to $\sigma_C = 14.7$ and $\sigma_E = 11.5$. Proportions of children with clinical levels are $p_C = 0.53$ and $p_E = 0.15$ in the experimental and control condition, corresponding to an estimated effect size $p_C - p_E = 0.38$ (Step 4).

The psychologist wishes to detect an intervention effect with high probability and selects a power level $1 - \beta = 0.90$. The probability of incorrectly rejecting the null hypothesis of no intervention effect is set to $\alpha = 0.05$. Two-sided alternative hypotheses $H_a : \mu_C \neq \mu_E$ and $H_a : p_C \neq p_E$ are chosen (Step 5) as one does not want to exclude the possibility that the experimental condition performs worse than the control.

Figure 3.3 shows the power level as a function of the sample size per treatment condition (Step 6). Equations (3.4) and (3.13) were used to calculate power as a function of sample size for the continuous and dichotomous outcome variables. From Figure 3.3 it follows that 121 mothers per condition are required for the continuous outcome, whereas only 28 mothers are needed for the dichotomous outcome (Step 7). Thus, the required sample size for a continuous outcome is different from the required sample size for a dichotomous outcome, although it should be strongly emphasized that it is not commonly the case that fewer subjects are required for a dichotomous outcome. This special case arises because the effect for the dichotomous outcome is much stronger than for the continuous outcome.

3.7 The optimal design of experiments

The power of a trial depends on the chosen research design. An optimal design uses the available resources in such a way that the parameters of the model used to relate treatment condition and relevant covariates to response are estimated with highest efficiency. Optimal designs in the field of toxicology provide the optimal number of dose levels, the optimal placement of dose levels among the dose axes and the optimal number of laboratory animals or

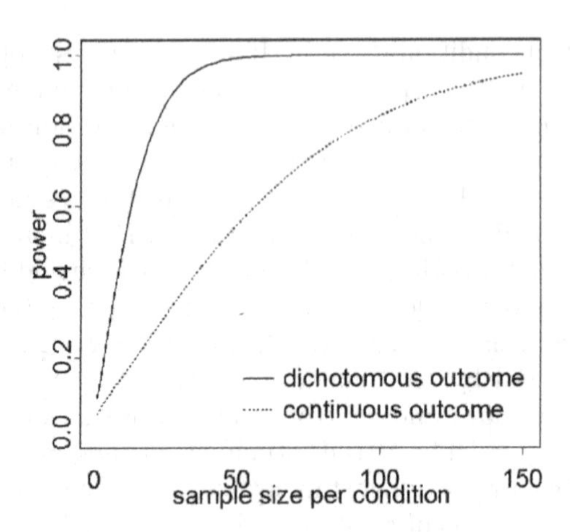

Figure 3.3: Power as a function of sample size per treatment condition.

patients per dose level. Optimal designs for cluster randomized trials provide the optimal number of clusters and optimal cluster size. Optimal designs for longitudinal studies provide the optimal number of measurement occasions and the optimal placement of these among the time axis. Fractional factorial designs study the optimal combinations of levels of various qualitative factors. A general introduction to optimal design theory can be found in Atkinson, Donev, and Tobias (2007), Berger and Wong (2009) and Goos and Jones (2011).

Optimal designs seek to maximize the information on the parameters of the model that relates outcome to treatment condition and covariates. This information is captured by the Fisher information matrix, of which the inverse is equal to the covariance matrix of the estimated model parameters. Choosing an optimal design means selecting a design among all possible designs that maximizes the Fisher information matrix. Since matrices cannot be ordered in a unique way, various functions have been proposed as optimality criteria and a design that is optimal given a particular criterion is not necessarily optimal for another criterion. The most common criterion is D-optimality, for which the determinant of the Fisher information matrix is maximized. Stated differently, a D-optimal design minimizes the volume of the confidence ellipsoid of the model parameters. A confidence ellipsoid can be regarded as a multidimensional confidence interval. Another approach is to minimize the variance of the estimator of one particular important model parameter, for instance, the treatment effect, as this is generally the parameter of main interest in experimental studies.

The performance of a sub-optimal design ξ relative to the optimal design ξ^* is generally expressed by means of a relative efficiency. The relative efficiency varies between zero and one; its inverse indicates the number of times the suboptimal design needs to be replicated to perform as well as the optimal design in terms of the chosen optimality criterion. Designs with efficiencies of 0.8 and 0.9 are generally preferred in practical situations since they do not perform much worse than the optimal design. A relative efficiency of 0.8 means that $(1/0.8 - 1)100\% = 25\%$ more individuals are needed to perform as well as the optimal design, while a relative efficiency of 0.9 means that sample size should be increased by $(1/0.9 - 1)100\% = 11\%$.

3.7.1 An example (continued)

The measurements on the externalizing disorder scale of the CBCL in the control and experimental groups have different variances σ_C^2 and σ_E^2 respectively. Also, the costs differ across the treatment conditions and are denoted c_C and c_E in the control and experimental conditions. If the variance of the treatment effect estimator is used as an optimality criterion, the optimal allocation ratio of the two sample sizes is

$$\frac{n_C}{n_E} = \sqrt{\frac{c_E \sigma_C^2}{c_C \sigma_E^2}}. \tag{3.21}$$

See Schouten (1999). This indicates that a higher sample size is required in the condition with more variable outcomes and lower costs. It follows that the optimal proportion of subjects in the control condition is equal to

$$p_C^* = \frac{n_C}{n_C + n_E} = \frac{\sqrt{c_E \sigma_C^2}}{\sqrt{c_E \sigma_C^2} + \sqrt{c_C \sigma_E^2}}. \tag{3.22}$$

Of course p_C^* lies in the range [0,1], which is called the design space. Figure 3.4 shows the relative efficiency of designs with $0 < p_C < 1$ in case the true standard deviations are equal to $\sigma_C = 14.7$ and $\sigma_E = 11.5$ and the costs are c_C=1 and $c_E = 5$. The efficiency is equal to 1 when $p_C = 0.74$ and decreases to lower values as the proportion p_C departs from $p_C = 0.74$. A design with equal allocation to treatment conditions (i.e., $p_C = 0.5$) has an efficiency of 0.78, meaning that $(1/0.78 - 1)100\% = 28\%$ more subjects are needed to perform as well as the optimal design.

3.8 Sample size and precision analysis

The goal of an a priori power analysis is to calculate the sample size such that a treatment effect can be detected at a certain probability. In addition to null hypothesis testing, a researcher might also wish to estimate the size of the treatment effect. The precision of such an estimate is expressed by a confidence interval. The precision of an estimate depends on the width of its confidence interval: the narrower the confidence interval, the more precise the

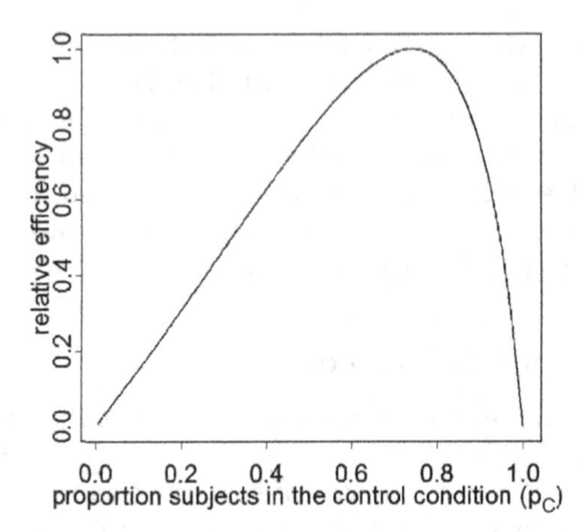

Figure 3.4: Relative efficiency as a function of the proportion of subjects in the control condition.

estimate. The upper and lower limits for a $(1-\alpha)100\%$ confidence interval are calculated from

$$\text{effect size} \pm z_{1-\alpha/2} \text{ standard error}, \tag{3.23}$$

which holds when the sample size is sufficiently large. Hypothesis testing and the confidence interval approach are related. The null hypothesis of no treatment effect is rejected when the confidence interval does not include the value zero. The lower the significance level α, the larger the width of the confidence interval and the more conservative the statistical test (i.e., the less often the null hypothesis is rejected).

The aim of a precision analysis is to calculate sample size such that the treatment effect is estimated at a certain precision, which is expressed in terms of the half-width d of confidence interval:

$$z_{1-\alpha/2} \cdot \text{ standard error} = d. \tag{3.24}$$

As for an a priori power analysis, the standard error depends on sample size and the required sample size can be calculated once the relation between standard error and sample size is known. As an example, consider a trial with two treatment conditions and continuous and uncorrelated outcomes. The standard error of the treatment effect estimator is $\sqrt{\sigma_C^2/n_C + \sigma_E^2/n_E}$, hence the half width of the $(1-\alpha)100\%$ confidence interval is

$$z_{1-\alpha/2}\sqrt{\frac{\sigma_C^2}{n_C} + \frac{\sigma_E^2}{n_E}}. \tag{3.25}$$

If the treatment groups are of equal size $n = n_C = n_E$, the required sample size per treatment condition for precision d is

$$n = \frac{z^2_{1-\alpha/2}(\sigma^2_C + \sigma^2_E)}{d^2}. \tag{3.26}$$

3.9 Sample size and accuracy of parameter estimates

Parameters of statistical regression models are generally estimated using maximum likelihood or restricted maximum likelihood, which are asymptotic in the sense that they require large sample sizes. In practice, sample sizes are limited and one may raise questions with respect to the acceptable lower limit to sample size. With hierarchical data, two sample sizes are involved: the number of groups and the group size. Increasing the number of groups is not an easy task if costs are involved, while sampling a subject in a group that is already included in a trial is often easier to perform. In addition, increasing the number of groups is not possible if all groups are already included or if the number of eligible groups is limited.

The effect of sample size on the accuracy of parameter estimates is often studied on basis of a simulation study, of which the earliest were published in the mid 1990s (Afshartous, 1995; Mok, 1996). A more recent paper by Maas and Hox (2005) used three different group sizes (5, 30 and 50) and three different numbers of groups (30, 50, 100). They conclude that standard errors of variance components at the group level are underestimated if the number of groups is 50 or less. For larger number of groups and all group sizes, the point estimates of regression coefficients and variance components and their standard errors were unbiased and accurate. When using Bayesian methods for data analysis, a smaller number of 20 groups is sufficient (J. Hox, Van De Schoot, & Matthijsse, 2012; J. J. Hox, Moerbeek, Kluytmans, & Van De Schoot, 2014).

Less promising results are found for multilevel logistic models. Even for large numbers of groups, the point estimates of the variance components in a multilevel logistic model may be subject to severe downward bias. This bias depends on the chosen estimation method.

Rodríguez and Goldman (1995) used a simulation study with 2449 births nested within 1558 mothers nested within 161 communities in Guatemala to study the performance of first order MQL. They reported biases up to 90% and in some cases the variance components were estimated to be equal to zero in all simulated data sets. The standard errors of the variance components were in general quite accurate. In addition, the point estimates of the fixed effects also showed a clear downward bias.

The same simulated data were used to study the performance of second order PQL (Rodríguez & Goldman, 2001; Goldstein & Rasbash, 1996). Point estimates of fixed effects were downward biased with at most 3% while point estimates of variance components showed a downward bias of at most 20%. For none of the simulated data sets were the variance components estimated equal to zero.

Another simulation study for multilevel logistic models was conducted by Moineddin, Matheson, and Glazieer (2007). They used the SAS procedure NLMIXED, which allows full maximum likelihood estimation. The authors recommend using a group size of at least 50 with at least 50 groups.

Moerbeek et al. (2003b) studied the performance of PQL in cluster randomized trials and multisite trials with dichotomous outcomes. They restricted their study to the treatment effect, which is the most important parameter in experiments, and used three different combinations of group size n_1 and number of groups n_2 that all resulted in a total sample size of 400: $(n_1, n_2) = (10, 40), (20, 20), (40, 10)$. They observed that non-convergence occurred up to 27.5% for multisite trials with site by treatment interaction and up to 5% for multisite trials without such interaction and for cluster randomized trials. The bias of the point estimate of the treatment effect was seldom larger than 10% and 5% for first and second order PQL, and the standard error of the treatment effect estimator was found to be almost unbiased.

As for all multivariate techniques, multilevel analysis requires large sample sizes. Especially small sample sizes at the group level may result in inaccurate point estimates of model parameters and their standard errors, especially those of variance components. Snijders and Bosker (2012) suggest 20 clusters may represent a random sample from the underlying population, but with such a small number of clusters the risk of inaccurate estimates is large.

4

Cluster randomized trials

4.1 Introduction

With cluster randomized trials, complete groups of subjects are randomized to treatments and all members of a given group receive the same treatment condition. Such trials are variously known as group randomized trials, place-based trials and community intervention trials. They are conducted in many disciplines in the social and medical sciences and examples are trials in general practice with patients nested within practices, trials in organizational studies with employees nested within worksites and school-based substance abuse prevention and cessation trials with pupils nested within schools.

Cluster randomized trials are less efficient than simple randomized trials where clustering does not occur. This is explained by the fact that measurements on outcomes on subjects within the same cluster are likely to be correlated. Reasons for such correlation were already presented in Section 1.2 and include therapist effects, mutual influence of group members and genetic factors. The degree of this correlation is expressed by the intraclass correlation coefficient, which can be regarded as the correlation between outcomes of two randomly selected subjects within the same cluster. The larger the intraclass correlation coefficient, the lower the efficiency of a cluster randomized trial relative to a simple randomized trial.

Given the loss of efficiency due to cluster randomization, an obvious question is why cluster randomized trials have become so popular over the past decades. Reasons for this are often of a political, ethical and financial nature rather than from a statistical basis (Gail, Mark, Carroll, & Green, 1996). Consider for instance a school-based substance abuse prevention intervention where pupils in the intervention condition receive lessons delivered by a health professional and pupils in the control condition do not receive any treatment. A large reduction in the health professional's travel costs may be achieved if randomization is done at the school level and only half of the schools are assigned to the intervention condition. A similar example is a clinical trial that requires expensive equipment in the experimental condition. A substantial reduction in costs may be achieved if this equipment is only available in half of the clinics.

One case where ethical and political reasons play a role in the choice of design is a trial on the effect of vitamin A supplementation on child mortality in rural villages in developmental countries. In such a trial it may be considered unacceptable to treat some children in a village and refrain others from treating. A cluster randomized trial with randomization at the village level is then preferred above a simple randomized trial. In community intervention trials, the intervention is designed to affect all members within the community using mass media, such as television and advertisements in newspapers, and there is no alternative to cluster randomization. A final reason to adopt a cluster randomized trial is the need to avoid control group contamination, which may occur when information on the intervention leaks to subjects assigned to the control as a result of direct communication. This is less likely when all members within a group receive the same treatment condition.

Cluster randomized trials appear to have many advantages; it is however necessary to be wary of the potential methodological bias of cluster randomization. Hahn, Puffer, Torgerson, and Watson (2005), for instance, compared the age distribution of subjects in a simple and cluster randomized trials on hip protectors and showed a tendency for cluster randomized trials to show an age imbalance between treatment groups. They also noted that cluster randomized trials showed a large positive treatment effect while the effects obtained with individually randomized trials ranged from negative to positive. The authors advocate individual randomization rather than cluster randomization whenever possible and identifying participants prior to allocation.

If such prior identification is not possible, they suggest using an independent recruiter to avoid selection bias. These findings, however, were not supported by a meta-analysis of enhanced care for depression conducted by Gilbody, Bower, Torgerson, and Richards (2008). They found no imbalance with respect to pretest depression scores in individual and cluster randomized trials. In addition, they found that cluster randomized trials produced almost identical effect size estimates as individually randomized trials. They concluded that the use of a cluster randomized trial needs to be justified if individually randomized trials produce robust and believable estimates.

The popularity of cluster randomized trials is stressed by the dedication of five books to this design (M. J. Campbell & Walters, 2014; Donner & Klar, 2000b; S. Eldridge & Kerry, 2012; Hayes & Moulton, 2009; Murray, 1998) and special issues in three major statistical journals (M. K. Campbell, Mollison, & Grimshaw, 2001; Donner & Klar, 2000a; Moulton, 2005). In addition, a number of review papers on this design have been published in the 2000s (M. J. Campbell, Donner, & Klar, 2007; Klar & Donner, 2001; Murray, Varnell, & Blitstein, 2004; Varnell, Murray, Janega, & Blitstein, 2004).

Donner, Klar, and Zou (2004) remark that the ethical principles for individually randomized trials may require modifications of cluster randomized trials. The guidelines in the Consolidated Standards of Reporting Trials (CONSORT) statement (Moher, Schulz, & Altman, 2001; K. F. Schulz, Altman, Moher, & Fergusson, 2011) include a checklist of items that should be included in trial reports. This statement was extended to cluster randomized trials in the mid 2000s (M. K. Campbell, Elbourne, & Altman, 2004).

Statistical papers on cluster randomization date back to the late 1970s. Cornfield (1978) already warned that "randomization by cluster accompanied by an analysis appropriate to randomization by individual is an exercise in self-deception, however, and should be discouraged."

Sample size requirements for cluster randomized trials became available in the early 1980s (Donner, Birkett, & Buck, 1981). One who expects the scientific community to be aware to use appropriate sample size calculations and analysis methods, however, is mistaken. Even in recent years, cluster randomization is not always correctly accounted for in the planning of a trial and the data analysis (Bland, 2004; M. K. Campbell, Grimshaw, & Elbourne, 2004; Crespi, Maxwell, & Wu, 2011; S. M. Eldridge, Ashby, Feder, Rudnicka, & Ukoumunne, 2004; Ivers et al., 2011; Murray, Pals, Blitstein, Alfano, & Lehman, 2008; Walleser, Hill, & Bero, 2011).

As the use of multilevel models for the analysis of data from cluster randomization trials is still not common practice, an introduction to this model is included in the next section and it is related to the analysis of variance model. Sample size calculations for continuous and dichotomous outcomes are presented in Sections 4.3 and 4.4, respectively. The focus is on sample sizes where the number of clusters or the cluster size is fixed, and sample sizes in trials with budgetary constraints are also presented. The last section shows how to use the sample size formulae in the design of a realistic trial. A restriction is made to study cluster randomized trials with two levels of nesting; the extension to three levels is made in Chapter 10.

4.2 Multilevel model

The multilevel model is introduced on the basis of a school-based smoking prevention intervention. Randomization to the intervention or control condition is done at the school level so all pupils within the same school receive the same treatment. The number of schools is denoted n_2 and the number of pupils in school j is denoted n_{1j}. For simplicity, it is assumed all schools are of equal size n_1, and the design is balanced with $\frac{1}{2}n_2$ schools per treatment condition. Artificial data of such an intervention with 20 schools of size 5 are displayed in Table 4.1. This table clearly illustrates that schools are nested within treatment conditions as each school participates in just one treatment. This design is called a nested design. The outcomes of interest are posttest scores on a test with items on smoking and health; test scores lie in the interval $[0, 100]$. Overall, higher scores are observed in the intervention condition, which suggests that pupils in the intervention condition are more aware of the negative health effects of smoking.

The pupil level regression model that models test score y_{ij} of pupil i in

Table 4.1: Example data for cluster randomized trial

school number	control					school number	experimental				
1	44	58	68	70	28	11	57	67	82	82	60
2	45	39	27	37	26	12	73	65	61	59	56
3	64	65	54	47	60	13	58	80	50	53	38
4	76	57	94	48	47	14	61	72	59	76	79
5	66	49	49	54	41	15	49	89	69	76	82
6	61	88	51	52	49	16	49	57	59	53	77
7	48	69	56	70	58	17	100	63	41	56	60
8	67	51	82	42	63	18	54	82	39	34	21
9	80	66	66	51	74	19	76	55	59	64	74
10	63	60	44	55	43	20	52	68	54	60	95

school j is given by

$$y_{ij} = \beta_{0j} + e_{ij}. \tag{4.1}$$

Each school j has its own mean score β_{0j} and the error term e_{ij} is the pupil level discrepancy between the observed score y_{ij} and the school's mean score β_{0j}. This mean score is predicted from treatment condition x_j, which gives the school level model

$$\beta_{0j} = \gamma_0 + \gamma_1 x_j + u_j. \tag{4.2}$$

Combining the two gives the single equation model

$$y_{ij} = \gamma_0 + \gamma_1 x_j + u_j + e_{ij}, \tag{4.3}$$

where γ_0 and γ_1 are the mean intercept and slope, respectively. Treatment condition is coded -0.5 for the control and 0.5 for the intervention, so that γ_0 is the grand mean and γ_1 is the difference in means in the two treatment groups. In other words, γ_1 is equal to the unstandardized treatment effect. Treatment condition has subscript j but not i since treatment varies over the schools j but not over the pupils i within a school. The random effects $u_j \sim N(0, \sigma_u^2)$ and $e_{ij} \sim N(0, \sigma_e^2)$ at the school and pupil level are assumed to be independent of each other.

Fitting model (4.3) to the data in Table 4.1 gives the results shown in Table 4.2. The null hypothesis of no treatment effect, $H_0 : \gamma_1 = 0$, is tested by the Wald test. The test statistic $t = \hat{\gamma}_1 / s.e.(\hat{\gamma}_1)$ follows a t distribution with $n_2 - 2$ degrees of freedom under the null hypothesis if REML estimation is used. The null hypothesis of no treatment effect is not rejected at the $\alpha = 0.05$ level. The intraclass correlation coefficient is calculated as $\rho = \sigma_u^2/(\sigma_u^2 + \sigma_e^2) = 33/(33 + 200) = 0.142$.

The estimated variance component at the school level is more than twice as large as its standard error, which suggests a significant part of the variance is located at the school level. The deviance test is a better test for variance components than the Wald test; see Section 2.2. For the multilevel model in

Table 4.2: Results based on the multilevel model

parameter	estimate	standard error	t	p
fixed effects:				
intercept, γ_0	59.67	1.916		
treatment effect, γ_1	6.46	3.832	1.686	0.109
random effects:				
intercept, σ_u^2	33.35	25.281		
error term, σ_e^2	200.36	31.680		

Table 4.2, the deviance is equal to 816.247. A deviance of 819.288 is obtained for the model that ignores the nesting of pupils within schools. These deviances are calculated on basis of restricted maximum likelihood, which is justified since the two models to be compared have the same fixed parts, and only differ in their random parts. The difference in deviances is 3.041 and follows a chi-squared distribution with 1 degree of freedom under the null hypothesis that the school level variance is equal to zero. The (halved) p-value is 0.041, thus the school level variance is significant at the 0.05 level.

For the current data set an incorrect conclusion with respect to the effect of treatment is achieved on basis of a regression model that ignores the nesting of pupils in schools. Such a model results in the same point estimate for treatment effect, but a much smaller standard error of 3.040. The corresponding p-value is equal to 0.036 and the null hypothesis of no treatment effect is rejected. This example therefore stresses the necessity to use a correct statistical model if a multilevel data structure is present in the data at hand.

The multilevel model (4.3) can be shown to be equivalent to a mixed effects nested ANOVA model (Raudenbush, 1993). In other words, the results of the multilevel model are mathematically identical to the results of the mixed effects ANOVA but they are presented in different forms. After all, the ANOVA model is just a special case of the regression model with categorical predictor variables that are generated by dummy variables. The ANOVA model that relates test score y_{ijk} of pupil k in school j in treatment condition i is

$$y_{ijk} = \mu + \alpha_i + \beta_{ij} + r_{ijk}, \tag{4.4}$$

where μ is the grand mean and α_i is the fixed effect associated with treatment i. To be able to estimate the effect of treatment the constraint $\sum_i \alpha_i = 0$ has to be posed. The random effects β_{ij} and r_{ijk} at the school and pupil level are assumed to be independently and normally distributed with zero mean and variances σ_β^2 and σ_r^2, respectively. There are $\frac{1}{2}n_2$ schools per treatment condition and n_1 pupils per school. If $i = 1$ for the control condition and $i = 2$ for the intervention condition, the correspondence between the mixed effects ANOVA model (4.4) and the multilevel regression model (4.3) is

$$\mu = \gamma_0,\ \alpha_2 - \alpha_1 = \gamma_1,\ \beta_{ij} = u_j,\ r_{ijk} = e_{ij}. \tag{4.5}$$

The degrees of freedom, sums of squares and expected mean squares for the ANOVA model are given in Table 4.3. The variance components are estimated from the mean squares as

$$\hat{\sigma}_\beta^2 = \frac{(MS_{school} - MS_{error})}{n_1} \text{ and } \hat{\sigma}_r^2 = MS_{error}. \tag{4.6}$$

The estimates are replaced by the value 0 if they are negative. The test statistic for the null hypothesis of no treatment effect is given by $F = MS_{treatment}/MS_{school}$ which, under the null hypothesis, has an F distribution with 1 and $n_2 - 2$ degrees of freedom.

Table 4.3: Analysis of a nested mixed effects ANOVA model

Source	*df*	*SS*	*E(MS)*
treatment	1	$\frac{1}{2}n_1 n_2 \sum_i (\overline{y}_{i..} - \overline{y}_{...})^2$	$\sigma_r^2 + n_1\sigma_\beta^2 + \frac{1}{2}n_1 n_2 \Sigma_i \alpha_i^2$
schools within treatment	$n_2 - 2$	$n_1 \sum_i \sum_j (\overline{y}_{ij.} - \overline{y}_{i..})^2$	$\sigma_r^2 + n_1\sigma_\beta^2$
error	$n_1 n_2 - n_2$	$\sum_i \sum_j \sum_k (y_{ijk} - \overline{y}_{ij.})^2$	σ_r^2
corrected total	$n_1 n_2 - 1$	$\sum_i \sum_j \sum_k (y_{ijt} - \overline{y}_{...})^2$	

Table 4.4 shows the results for the mixed effects ANOVA model on basis of REML estimation. The variance components at the school and pupil level are estimated from Equation (4.6) and are equal to $\hat{\sigma}_\beta^2 = 200.360$ and $\hat{\sigma}_r^2 = 33.35$, and these values correspond to the estimated variance components for the multilevel model. In addition, the estimates $\hat{\mu} = \bar{y}_{...} = 59.67$ and $(\hat{\alpha}_2 - \hat{\alpha}_1) = (\bar{y}_{2..} - \bar{y}_{1..}) = 6.46$ correspond to the estimated γ_0 and γ_1 in the multilevel model. The F test statistic for the ANOVA model is equal to the square of the t test statistic for the multilevel model, and for both models the p-value for the test on treatment effect is equal to 0.109.

Although mixed effects models have been widely used in the past, it appears that the multilevel model has become more popular nowadays. This may be explained by the fact that it can deal with varying school sizes and continuous covariates more easily. In the remainder of this chapter, power calculations for models with treatment condition as the sole predictor variable are given; the inclusion of covariates is discussed in the next chapter.

4.3 Sample size calculations for continuous outcomes

In this section, sample size formulae for trials with continuous outcomes are studied, for example, sum scores on questionnaires and measurements on physiological parameters. The factors that influence power are studied first: number of clusters, cluster size, intraclass correlation coefficient and effect size. Subsequently the focus is on the design effect that compares the efficiency

Table 4.4: Results based on the mixed effects ANOVA model

Source	*df*	*SS*	*MS*	*F*	*p*
treatment	1	1043.290	1043.290	2.842	0.109
schools within treatment	18	6608.020	367.112	1.832	0.035
error	80	16028.800	120.000		
corrected total	99	23680.110			

of a cluster randomized trial to that of a simple randomized trial. Next, the required number of clusters when cluster size is fixed and the required cluster size when the number of clusters is fixed are given. Simple closed form equations for these sample sizes can be obtained when cluster sizes do not vary and both treatment groups have equal numbers of clusters, but guidelines on sample size for trials where these restrictions are relaxed are also given. Thereafter, a cost function is used as precondition to derive sample size formulae for trials where neither cluster size nor number of clusters is fixed. Such a cost function specifies the costs for sampling a cluster and recruiting a subject in an already sampled cluster and allows us to calculate the optimal combination of cluster size and number of clusters such that the non-overhead costs are minimized.

4.3.1 Factors that influence power

As explained in the previous section, the test statistic for the test on treatment effect follows a central t distribution with $n_2 - 2$ degrees of freedom in case the null hypothesis is true. Under the alternative hypothesis this test statistic follows a non-central t distribution with non-centrality parameter

$$\lambda = \frac{\gamma_1}{\sqrt{\frac{4(\sigma_e^2 + n_1\sigma_u^2)}{n_1 n_2}}}. \tag{4.7}$$

The numerator of this ratio is the unstandardized effect size; the denominator is the standard error of its estimator

$$s.e.(\hat{\gamma}_1) = \sqrt{\frac{4(\sigma_e^2 + n_1\sigma_u^2)}{n_1 n_2}}; \tag{4.8}$$

see Raudenbush (1997) and Moerbeek, Van Breukelen, and Berger (2000, 2003c). The power for the test is calculated as

$$\begin{array}{r} P(t_{n_2-2,\lambda} > t_{n_2-2,1-\alpha}) \text{ when } H_a : \gamma_1 > 0 \\ P(t_{n_2-2,\lambda} < t_{n_2-2,\alpha}) \text{ when } H_a : \gamma_1 < 0 \\ P(t_{n_2-2,\lambda} > t_{n_2-2,1-\alpha/2}) + P(t_{n_2-2,\lambda} < t_{n_2-2,\alpha/2}) \text{ when } H_a : \gamma_1 \neq 0 \end{array} . \tag{4.9}$$

As was noted in the previous chapter, power increases with increasing type I error rates α and the effects of α are not studied further in this or subsequent chapters. The graphs in this section are based on a test with a two-sided alternative hypothesis and $\alpha = 0.05$.

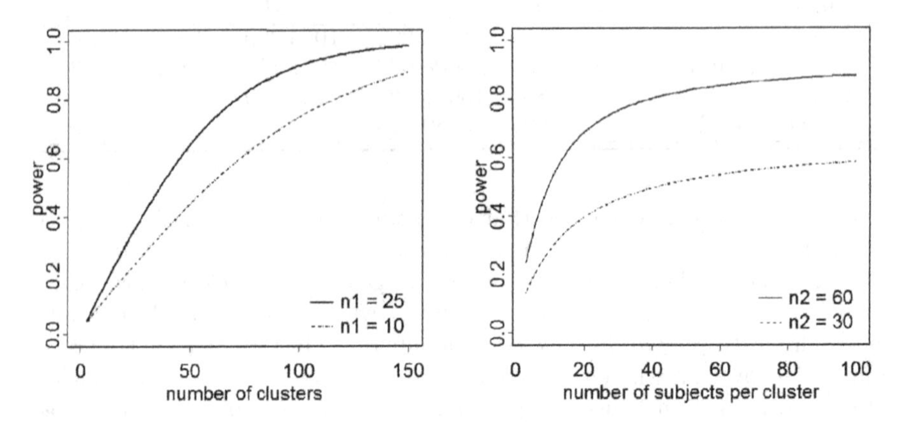

Figure 4.1: Power as a function of the number of clusters and cluster size.

The effect size, standard error and non-centrality parameter are related to the power of the test on treatment effect: the larger the effect size and the smaller its standard error, the larger the non-centrality parameter, and hence the larger the power. It is observed that the non-centrality parameter depends on the number of clusters n_2, the cluster size n_1, the variance components σ_e^2 and σ_u^2 and the unstandardized effect size γ_1. The effects of these factors are now discussed.

Number of clusters. From Equation (4.8) it is observed that the number of clusters appears in the denominator of the standard error but not in the numerator. As a consequence, the standard error approaches zero, and hence the power approaches the value 1 if the number of clusters increases to infinity. This is illustrated in the left side of Figure 4.1, where the power to detect a small effect size in a trial with $\rho = 0.05$ is plotted as a function of the number of clusters. The two lines represent cluster sizes $n_1 = 10$ and $n_1 = 25$. Clearly, the power is larger if cluster size is larger, and for both cluster sizes the power approaches the value 1 if the number of clusters increases to sufficiently large values. In theory, one could reach a desired power level by increasing the number of clusters to a required level. In practice, however, this approach is not always feasible since the number of clusters that can be enrolled in a study may be restricted for various reasons. For instance, if enrolling a cluster is expensive, one may choose to increase power by recruiting more subjects per cluster.

Cluster size. The cluster size appears in both the numerator and denominator of the standard error of the treatment effect. Increasing the cluster size to infinity implies that the standard error approaches $\sqrt{4\sigma_u^2/n_2}$, which is only equal to zero if $\sigma_u^2 = 0$ (Guittet, Giraudeau, & Ravaud, 2005). That is, increasing cluster size to infinity only results in a power level of 100% if the between-cluster variability with respect to the outcome variable is equal to zero. Increasing a small cluster size will have some effect on power but at some point recruiting more subjects within a cluster hardly results in higher power levels. A higher power level can then be achieved by increasing the number of clusters or by matching, pre-stratification or the inclusion of covariates.

Thus, cluster size has a weaker effect on power than the number of clusters. This is illustrated in the right side of Figure 4.1 where power to detect a small treatment effect is plotted as a function of the cluster size, and for number of clusters $n_2 = 30$ and $n_2 = 60$. The intraclass correlation coefficient is set at $\rho = 0.05$. For small cluster sizes, increasing cluster size results in a substantial increase of the power level, but increasing cluster size above $n_1 = 40$ has only a small effect on power. A maximum power of about 58% can be achieved with a design with just $n_2 = 30$ clusters; increasing the number of clusters to $n_2 = 60$ results in a higher power level above the commonly desired 80% and 90% power levels.

Variance components and intraclass correlation coefficient. Increasing the variance components has a decreasing effect on power. The effect of increasing σ_u^2 is even amplified by the fact that σ_u^2 is multiplied by the cluster size n_1 in the standard error (4.8). Given fixed total variance $\sigma^2 = \sigma_e^2 + \sigma_u^2$ of the outcome y_{ij}, it is interesting to examine the effect of the intraclass correlation coefficient ρ. The standard error is reformulated as

$$s.e.(\hat{\gamma}_1) = \sqrt{\frac{4\sigma^2(1+(n_1-1)\rho)}{n_1 n_2}}. \tag{4.10}$$

For given group size n_1 and number of groups n_2, this standard error increases as the intraclass correlation coefficient ρ increases. This is illustrated in the left side of Figure 4.2, where power to detect a small, medium or large effect size in a test with a two-sided alternative hypothesis and $\alpha = 0.05$ is plotted as a function of the intraclass correlation coefficient.

The cluster size and number of clusters are equal to 30. Power decreases with increasing ρ and the value at which power drops most rapidly increases with increasing standardized effect size. From Equation (4.10) it follows that the effect of the intraclass correlation coefficient is stronger if cluster size is larger. So, for large intraclass correlation coefficients it is advisable to increase the number of clusters rather than cluster size to enhance power.

Effect size. Larger effect sizes are easier to detect and therefore the power increases with effect size γ_1. If a standardized effect size $\delta = \gamma_1/\sigma$ is used, the non-centrality parameter is rewritten as

$$\lambda = \frac{\delta}{\sqrt{\frac{4(1+(n_1-1)\rho)}{n_1 n_2}}}. \tag{4.11}$$

From this equation it follows that the total variance σ^2 does not have to be known to perform a power calculation. The right side of Figure 4.2 shows that power increases with increasing effect size and decreasing intraclass correlation coefficient. In this graph, the cluster size and number of clusters were equal to 30, but the same result is found for other sample sizes.

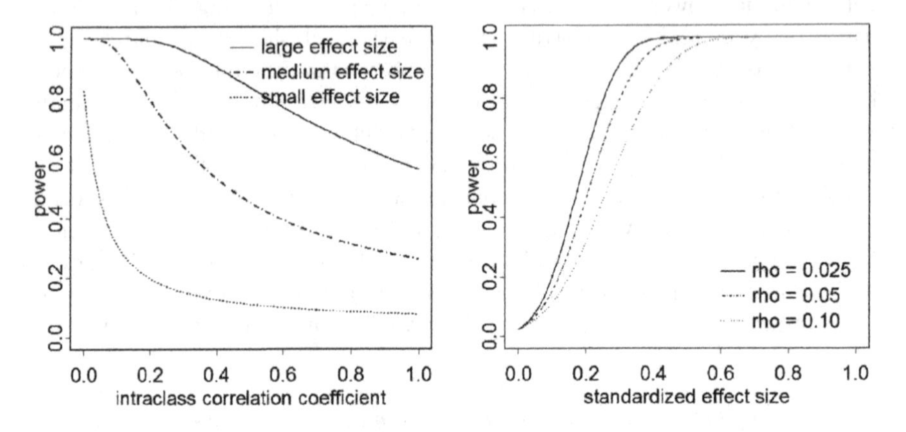

Figure 4.2: Power as a function of intraclass correlation coefficient and standardized effect size.

4.3.2 Design effect

With cluster randomized trials, complete clusters of subjects are randomized to treatment conditions. The outcomes of subjects within the same cluster are likely to be correlated; see Section 1.2 for possible reasons for such correlation. The degree of correlation is expressed by the intraclass correlation coefficient ρ. When $\rho = 1$, all subjects within a cluster respond identically; when $\rho = 0$ then the subjects within a cluster are no more correlated than subjects from different clusters. In general, the intraclass correlation coefficient lies somewhere in the interval $[0, 1]$ and the higher its value, the lower the efficiency of a cluster randomized trial compared to a simple randomized trial.

This efficiency is also referred to as the design effect and for cluster randomized trials is equal to

$$\text{design effect} = 1 + (n_1 - 1)\rho. \tag{4.12}$$

The design effect for cluster randomized trials is larger than 1, except for trials where $\rho = 0$ and/or $n_1 = 1$. This means that a cluster randomized trial requires a larger sample size to perform as well in terms of statistical power as a simple randomized trial. A two-step procedure can be used to calculate the required sample size for a cluster randomized trial. First, the required sample size for a simple randomized trial without clustering can be calculated on basis of Equation (3.5). This sample size is then multiplied by the design effect to obtain the required sample size for a cluster randomized trial. In the example in the previous chapter, a sample size of 121 mothers per condition was required for the continuous outcome variable externalizing disorder. A parallel group design was used in this example and mothers were randomly assigned to treatment conditions. It could well have been that mothers were nested within clusters, such as health care centers or shelter homes, and randomization was done at the cluster level. A larger sample size would then have been needed. If the common cluster size is $n_1 = 10$ and the intraclass correlation is $\rho = 0.05$, the design effect is 1.45, meaning that 1.45 as many subjects are needed in a cluster randomized trial than in a trial without nesting of subjects within clusters.

It should be noted that the design effect of a cluster randomized trial is equal to the design effect of a simple random sample to a two-stage sample; see Section 2.3. Figure 2.3 presents this design effect as a function of cluster size and intraclass correlation coefficient.

4.3.3 Sample size formulae for fixed cluster size or fixed number of clusters

Once a minimal clinically relevant effect size has been chosen and a prior estimate of the intraclass correlation coefficient has been made, the required values of the cluster size n_1 and number of clusters n_2 to achieve a desired power level can be calculated such that the desired power level is achieved. Various combinations (n_1, n_2) result in the desired power level. For instance, the combinations (10,116) and (25,72) both result in a power of 80% to detect a small treatment effect in a test with a two-sided alternative with $\alpha = 0.05$ and $\rho = 0.05$, as can be observed from the left side of Figure 4.1. The choice of the best combination (n_1, n_2) may then be driven by practical considerations.

In some trials, the number of subjects per cluster is fixed and the required number of clusters needs to be calculated. This may be the case in school-based interventions where class sizes are fixed or in therapy trials where the number of clients that can be treated by a therapist is fixed. An explicit formula for n_2 cannot be derived on basis of Equation (4.9), as the number of clusters appears in both the degrees of freedom and the non-centrality parameter of the t distribution. For large degrees of freedom, however, the t distribution can be approximated by the standard normal. As a rule of thumb, the normal approximation can be used when $df = 30$, thus when $n_2 = 32$ since $df = n_2 - 2$. The minimal required number of clusters is then calculated from

$$\begin{aligned} n_2 &= 4\frac{\sigma_e^2 + n_1\sigma_u^2}{n_1}\left(\frac{z_{1-\alpha} + z_{1-\beta}}{\gamma_1}\right)^2 \\ &= 4\frac{1 + (n_1 - 1)\rho}{n_1}\left(\frac{z_{1-\alpha} + z_{1-\beta}}{\delta}\right)^2. \end{aligned} \tag{4.13}$$

The sample size as calculated from this formula is often a real value and needs to be rounded upward to the nearest even integer to assure the desired power level is achieved. Rounding is done to the nearest even integer because both treatment arms are assumed to have equal number of clusters.

In other trials the number of clusters willing to participate is limited and the required cluster size is calculated from

$$n_1 = \frac{4\sigma_e^2}{\left(\frac{\gamma_1}{z_{1-\alpha}+z_{1-\beta}}\right)^2 n_2 - 4\sigma_u^2} = \frac{4(1-\rho)}{\left(\frac{\delta}{z_{1-\alpha}+z_{1-\beta}}\right)^2 n_2 - 4\rho}. \tag{4.14}$$

Again, rounding upward to the nearest integer is required. Equations (4.13) and (4.14) hold for a one-sided alternative hypothesis and for a two-sided alternative α needs to be replaced by $\alpha/2$.

As previously shown in this chapter, a trial with a limited number of clusters will not always be able to achieve a desired power level, even when it is possible to increase the cluster size (Guittet et al., 2005). Hemming, Girling, Sitch, Marsh, and Lilford (2011) provide a feasibility check to determine whether a sufficient number of clusters is available. This will occur when $n_2 > 2n\rho$, where n is the group size for a simple randomized trial as calculated from Equation (3.5). When the design is not feasible, one should decrease the desired power level and/or increase the effect size.

Equation (4.13) holds when cluster sizes do not vary. A design with equal cluster sizes can be shown to be more efficient than a design with varying cluster sizes. That is, fewer clusters are required to achieve a certain power when cluster sizes are fixed than when they vary. In practice, however, the clusters in a trial do not have equal size and an approximation to the required number of clusters can be obtained by replacing n_1 by the average cluster size.

This approximation does not take into account the imbalance in cluster size, and this imbalance can have a large effect, especially if the number of clusters is low and/or the intraclass correlation coefficient is high (Guittet, Ravaud, & Giraudeau, 2006).

Hoover (2002) provides SAS code to perform power calculations based on an approximation to the t distribution for trials with varying cluster sizes. Alternatively, the required number of clusters may be calculated based on the relative efficiency of designs with unequal versus equal cluster sizes. A lower bound of this relative efficiency is given by Manatunga, Hudges, and Chen (2001). This lower bound performs well when ρ is close to zero, but should not be used when $\overline{n}_1\rho > 1$. Van Breukelen, Candel, and Berger (2007) derive an approximation to the relative efficiency and compare this with results from a simulation study. They take into account a range of cluster size distributions and conclude that the accuracy of their approximation is quite good. They find that the minimal relative efficiency is at least 0.90, and the loss in efficiency of using varying cluster sizes can be compensated by including 11% more clusters.

Table 4.5: Reduction in non-overhead costs using optimal allocation ratio r as a function of costs ratio c_E/c_C

cost ratio	1	2	3	4	5	6	7	8	9	10
cost reduction (%)	0	2.9	6.7	10.0	12.7	15.0	16.9	18.6	20.0	21.3

Unequal allocation ratios

Thus far, sample size formulae for trials with $\frac{1}{2}n_2$ clusters per treatment condition were derived. Dumville, Hahn, Miles, and Torgerson (2006) performed a systematic review of simple randomized trials that used unequal randomization ratios. In almost all trials with two treatment arms, patients were recruited in favor of the experimental group. Reasons for unequal randomization in favor for the experimental group were the gaining of additional information on the experimental treatment, ethics and avoiding loss of power from drop-out or crossover. The need to reduce costs and limited availability of the intervention were reasons for randomization in favor of the control group.

Design efficiency is another reason to use unequal randomization ratios; see Hsu (1994) and Section 3.7 of this book. Liu (2003) studies the case where the costs to enroll a cluster in the trial vary over the two treatment conditions, but the total variance, intraclass correlation coefficient and cluster size are constant over the treatment conditions. The non-overhead costs for a trial with n_{2_C} and n_{2_E} clusters in the control and experimental condition, and c_C and c_E costs per cluster in the control and experimental condition are equal to $n_{2_E}c_E + n_{2_C}c_C$. The optimal sample allocation ratio r for experimental and control clusters is $r = n_{2_E}/n_{2_C} = \sqrt{c_C/c_E}$. For both equal and unequal randomization ratios, the costs to achieve a desired power level can be calculated. The ratio of these costs is equal to

$$\frac{Costs_{unequal}}{Costs_{equal}} = \frac{(1+r)(c_E + c_C/r)}{2(c_C + c_E)}. \tag{4.15}$$

The percentage reduction in non-overhead costs using unequal sample allocation is equal to $100\%(1 - Costs_{unequal}/Costs_{equal})$ and is given in Table 4.5 for cost ratios up to $c_E/c_C = 10$. Higher reductions are achieved for higher cost ratios. As can be observed, substantial reductions of 10% and higher can be achieved when the cost ratio is 4 or larger.

4.3.4 Including budgetary constraints

Thus far, either the cluster size or the number of cluster was assumed fixed and the other sample size was calculated such that a desired power level was achieved. Larger sample size at either the cluster or subject level results in larger statistical power. In practice, limited financial resources prevent the sample sizes to increase without bound. The focus is now on sample size formulae where neither the cluster size nor the number of clusters is fixed and a budgetary constraint is used to calculate the optimal allocation of units (n_1, n_2).

The costs to recruit a cluster in a cluster randomized trial may well be different from the costs to sample a subject within an already recruited cluster. Let c_1 and c_2 denote the costs at the subject and cluster level. The non-overhead costs of the trial are calculated by multiplying the total number of subjects by the subject level costs and adding the product of the number of clusters and costs at the cluster level. These costs should not exceed the budget C that is available for including clusters and subjects:

$$c_1 n_1 n_2 + c_2 n_2 \leq C. \tag{4.16}$$

If the costs c_1 and c_2 vary over treatment conditions, the mean costs may be used. This is permitted if the design is balanced with non-varying cluster sizes and equal numbers of clusters per treatment condition. In the remainder of this section, the focus is on such balanced designs. The aim is to minimize the standard error (4.8) such that the budgetary constraint is fulfilled. This strategy results in highest possible power, given the available budget C. The following procedure is used to derive the optimal allocation of units. The budgetary constraint is used to express the optimal number of clusters in terms of the cluster size, costs and budget:

$$n_2 = C/(c_1 n_1 + c_2). \tag{4.17}$$

This expression is then substituted in the standard error (4.8) and the optimal cluster size n_1^* is derived. The optimal cluster size, in its turn, is substituted into expression (4.17) to derive the optimal number of clusters n_2^*. The optimal cluster size is equal to

$$n_1^* = \sqrt{\frac{c_2 \sigma_e^2}{c_1 \sigma_u^2}} = \sqrt{\frac{c_2(1-\rho)}{c_1 \rho}}, \tag{4.18}$$

and the optimal number of clusters is equal to

$$n_2^* = \frac{C}{\sqrt{\frac{c_1 c_2 \sigma_e^2}{\sigma_u^2}} + c_2} = \frac{C}{\sqrt{\frac{c_1 c_2 (1-\rho)}{\rho}} + c_2}. \tag{4.19}$$

Given these optimal sample sizes, the standard error of the treatment effect estimator is calculated from

$$s.e.(\hat{\gamma}_1) = 2\frac{\sqrt{c_1 \sigma_e^2} + \sqrt{c_2 \sigma_u^2}}{\sqrt{C}} = 2\sigma\frac{\sqrt{c_1(1-\rho)} + \sqrt{c_2 \rho}}{\sqrt{C}}. \tag{4.20}$$

This standard error can be imputed in Equations (4.7) and (4.9) to calculate the corresponding power. In most cases, the optimal sample sizes have to be rounded to integers in such a way that the budgetary constraint is still satisfied. The standard error then becomes somewhat larger than the one calculated from Equation (4.20) and hence power becomes somewhat smaller.

The optimal sample sizes depend on the costs, budget and intraclass correlation coefficient. Equation (4.18) shows that the optimal cluster size decreases as the intraclass correlation coefficient ρ increases. This is obvious since one would sample fewer subjects per cluster if the degree of dependency among subjects within a cluster is high. In the (theoretical) extreme case where $\rho = 1$, all subjects respond identically and sampling just one subject per cluster suffices. Furthermore, the optimal cluster size increases with increasing cost ratio c_2/c_1. This is also obvious, since one would sample more subjects per cluster instead of sampling more clusters when the costs at the cluster level are high relative to the costs at the subject level. Finally, changing the budget has an effect on the optimal number of clusters, but not on the optimal cluster size. Increasing the budget results in a lower $s.e.(\hat{\gamma}_1)$ as can be deduced from Equation (4.20), and hence a higher power level.

These findings are visualized in Figure 4.3 for a trial with $c_1 = 10$ and $c_2 = 262$ Dutch guilders, the monetary unit used in the Netherlands before the introduction of the Euro. These costs were found for a school-based smoking prevention intervention conducted in the Netherlands in the late 1990s (Ausems, Mesters, Van Breukelen, & De Vries, 2002; Moerbeek et al., 2003c). The intraclass correlation coefficient used for the calculation of the optimal sample sizes is set at $\rho = 0.07$. Figure 4.3 shows the power to detect a small standardized treatment effect in a test with a two-sided alternative hypothesis and $\alpha = 0.05$ as a function of cluster size and number of clusters for three different budgets: $C = 60000$, $C = 40000$, and $C = 20000$. From the left side of Figure 4.3, we see that the budget C does not have an effect on the optimal cluster size as for each budget the highest power is achieved if $n_1 = 18.7$. The right side of this figure shows that the budget does have an effect on the optimal number of clusters. Highest power is achieved when $n_2 = 133.8$ for $C = 60000$, when $n_2 = 89.2$ for $C = 40000$ and when $n_2 = 44.6$ for $C = 20000$. Furthermore, it is observed from both subfigures that power increases if a higher budget is used. For this example the budget to reach 80% power is somewhat larger than $C = 40000$.
Instead of maximizing power given a fixed budget, one could also minimize the budget to achieve a desired power level. For the latter approach, the optimal cluster size and number of clusters are still given by Equations (4.18) and (4.19), provided the *df* are large. The required budget can then be approximated by using the normal approximation to the t distribution:

$$C = \frac{4\sigma^2(\sqrt{c_1(1-\rho)} + \sqrt{c_2\rho})^2(z_{1-\alpha} + z_{1-\beta})^2}{\gamma_1^2}. \tag{4.21}$$

For a two-sided alternative hypothesis α needs to be replaced by $\alpha/2$. In practice, the budget will be somewhat larger due to rounding of the sample

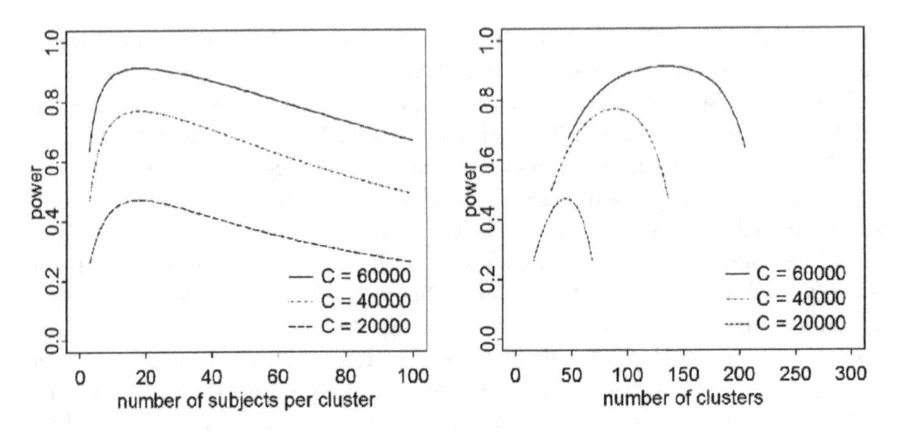

Figure 4.3: Power as a function of cluster size (left) and number of clusters (right) for a trial with a budgetary constraint.

sizes to integers. For small *df*, the power level is not only determined by the $s.e.(\hat{\gamma}_1)$ but also by the $df = n_2 - 2$ and Equations (4.18) and (4.19) do not necessarily result in the optimal design. In such cases one could find the optimal design by evaluating a large number of designs (n_1, n_2) and selecting the one for which the desired power level is achieved at lowest costs.

It should be observed that the optimal design as determined by the optimal cluster size n_1^* and number of clusters n_2^* depends on the intraclass correlation coefficient ρ. Such a design is called a locally optimal design in optimal design theory. A good prior estimate of ρ from the literature, expert knowledge or a pilot is required to calculate the optimal design. Korendijk, Moerbeek, and Maas (2010) studied the robustness of the optimal allocation against incorrect estimates of ρ for a range of plausible population values. They found that underestimation of ρ results in a steeper decrease of design efficiency than overestimation does. However, the design seems to be rather robust to incorrect prior estimates as design efficiency remains above 90% as long as the intraclass correlation coefficient is not underestimated by more than 75% or overestimated by more than 175%.

4.4 Sample size calculations for dichotomous outcomes

Sample size formulae for cluster randomized trials with nested data and dichotomous outcomes are extensions of those presented in Section 3.1.3 and show similarities to those for continuous outcomes in the previous section. Sample size formulae based on the risk difference and the odds ratio are given.

In both approaches, p_C and p_E are the probabilities of event occurrence in the control and experimental condition, respectively. In the calculations that follow, a one-sided alternative hypothesis is used. For two-sided alternative hypotheses, α should be replaced by $\alpha/2$.

4.4.1 Risk difference

The null hypothesis states zero risk difference between the control and experimental condition $H_0 : p_C - p_E = 0$. The risk difference is estimated by $\hat{p}_C - \hat{p}_E$, which has standard error

$$s.e.(\hat{p}_C - \hat{p}_E) = \sqrt{\left(\frac{p_C(1-p_C)}{\frac{1}{2}n_1 n_2} + \frac{p_E(1-p_E)}{\frac{1}{2}n_1 n_2}\right)(1+(n_1-1)\rho)}. \quad (4.22)$$

The intraclass correlation coefficient is calculated on basis of mean squares, as explained in Section 2.4. The standard error is estimated by replacing the proportions by their estimates $\hat{p}_C$ and $\hat{p}_E$. For large sample sizes, the test statistic $(\hat{p}_C - \hat{p}_E)/\hat{s.e.}(\hat{p}_C - \hat{p}_E)$ follows a standard normal distribution under the null hypothesis. The required number of clusters given a fixed cluster size n_1 is

$$n_2 = \frac{(p_C(1-p_C) + p_E(1-p_E))(1+(n_1-1)\rho)}{\frac{1}{2}n_1}\left(\frac{z_{1-\alpha} + z_{1-\beta}}{p_C - p_E}\right)^2. \quad (4.23)$$

As for continuous outcomes, the required number of clusters increases with the design effect $1 + (n_1 - 1)\rho$.

Given a fixed number of clusters n_2, the required cluster size is

$$n_1 = \frac{(1-\rho)(p_C(1-p_C) + p_E(1-p_E))}{\left(\frac{p_C - p_E}{z_{1-\alpha}+z_{1-\beta}}\right)^2 \frac{1}{2}n_2 - \rho(p_C(1-p_C) + p_E(1-p_E))}; \quad (4.24)$$

see for instance Donner and Klar (2000b). Equation (4.24) was further studied by Ahn, Hu, Skinner, and Ahn (2009). They concluded that the power levels obtained from this formula are close to the nominal levels obtained from a simulation study, even when the number of clusters per treatment condition is as small as 10. So, even for trials with small number of clusters Formula (4.24) provides a good approximation to the nominal power.

In case neither the cluster size nor the number of clusters is fixed, the budgetary constraint (4.16) can be used to calculate the optimal cluster size n_1^* and number of clusters n_2^*. These optimal sample sizes follow from the equations at the right of Formulae (4.18) and (4.19). Thus, the optimal allocation of units, given a budget C, depends on the intraclass correlation coefficient and costs at the subject and cluster level, but not on the event probabilities in both treatment arms.

4.4.2 Odds ratio

The probability p_{ij} of an outcome $y_{ij} = 1$ for subject i in cluster j is modeled as

$$p_{ij} = \frac{1}{1 + e^{-(\gamma_0 + \gamma_1 x_j + u_j)}}, \tag{4.25}$$

where the error u_j at the cluster level is assumed to follow a normal distribution with zero mean and variance σ_u^2. As previously, treatment condition x_{ij} is coded -0.5 and 0.5 for the control and intervention groups. The treatment effect estimator $\hat{\gamma}_1$ is the log odds ratio and has standard error

$$s.e.(\hat{\gamma}_1) = \sqrt{\frac{4(\sigma^2 + n_1\sigma_u^2)}{n_1 n_2}}, \tag{4.26}$$

where

$$\sigma^2 = \frac{1}{2}\left(\frac{1}{p_C(1-p_C)} + \frac{1}{p_E(1-p_E)}\right) \tag{4.27}$$

quantifies the variability at the subject level (Moerbeek, Van Breukelen, & Berger, 2001a). The standard error (4.26) for a dichotomous outcome is obtained from the standard error (4.8) for a continuous outcome by replacing the subject level variance component σ_e^2 by its counterpart from (4.27). This implies that results from the previous section for continuous outcomes also apply to cluster randomized trials with dichotomous outcomes. The power decreases with increasing between-cluster variability in outcomes, and increases with increasing effect size, cluster size and number of clusters. Once again, the effect of the number of clusters is larger than the effect of cluster size. Using small clusters can be compensated by sampling many clusters but using few clusters cannot always be compensated by sampling many subjects per cluster.

Asymptotically, the test statistic $\hat{\gamma}_1/s.e.(\hat{\gamma}_1)$ follows a standard normal distribution under the null hypothesis of no treatment effect. The sample size formulae for continuous outcomes can be used for dichotomous outcomes once the subject level error term σ_e^2 is replaced by σ^2 from Equation (4.27). These formulae are not repeated here.

It should be noted that Formulae (4.26) and (4.27) were obtained with the first order MQL estimation method, while it is well known that second order PQL performs better. The standard error as obtained from these formulae needs to be multiplied by a conversion factor of about 1.1 to hold for second order PQL. This implies that sample sizes as calculated from the formulae in the previous section are slightly too low. The correct sample sizes should be calculated from the relation

$$1.1\sqrt{\frac{4(\sigma^2 + n_1\sigma_u^2)}{n_1 n_2}} = \frac{\gamma_1}{z_{1-\alpha} + z_{1-\beta}}. \tag{4.28}$$

All sample size formulae in this section assume non-varying cluster sizes. Ahn, Hu, and Skinner (2009) provide an extension to Equation (4.23) for trials with unequal cluster sizes and show that this extension provides power levels that are closer to the nominal value than Equation (4.23) with the cluster size n_1 replaced by the mean cluster size $\overline{n}_1$. Their formula is similar to that in Manatunga et al. (2001) for continuous outcomes and works well when ρ is close to 0, but not so when $\overline{n}_1\rho > 1$. For unequal cluster sizes, Candel and Van Breukelen (2010) conclude that in many cases sampling 14% more clusters is sufficient to repair the efficiency loss due to varying cluster sizes. They also calculated a correction factor for using second order PQL instead of first order MQL, and it appeared to be of the same magnitude as the correction factor of about 1.1 for non-varying cluster sizes mentioned above.

4.5 An example

A large study evaluated the effect of a smoking prevention intervention aimed at pupils in the eighth grade of Dutch elementary schools (Ausems et al., 2002). The pupils were about 11 and 12 years of age and would transfer to a secondary school by the end of the school year. The study evaluated the effect of an existing in-school prevention program in which pupils received information on different aspects of smoking and health during class lessons. An out-of-school intervention in which pupils received tailored letters on smoking and health at their home addresses was evaluated simultaneously since previous research indicated that the in-school condition would function more effectively when supplemented with the out-of-school intervention. Results of the study showed positive effects regarding the out-of-school program but no significant effects were found for the in-school program. The combined approach did not show an improvement over the single-method approach.

Suppose a researcher wants to evaluate the effects of the out-of-school intervention in another country. We show how to perform sample size calculations for a continuous outcome. Sample size calculations for a dichotomous outcome are rather similar and not presented here. The continuous outcome of interest is a variable that measures the attitude toward the disadvantages of smoking. It is measured on basis of 11 items that range from very positive (1) to very negative (5) toward smoking, so the minimum and maximum sum scores are 11 and 55. Suppose the variances at the pupil and school level are equal to $\sigma_e^2 = 62$ and $\sigma_u^2 = 8$, which implies the total variance is $\sigma_e^2 + \sigma_u^2 = 70$ and the intraclass correlation coefficient is $\rho = 8/70 = 0.11$. The researcher wants to detect an unstandardized effect size $\gamma_1 = 2$, which corresponds to a standardized effect size $\delta = 2/\sqrt{70} = 0.24$, which is a small effect. The desired power level is $1 - \beta = 0.8$ at a significance level $\alpha = 0.05$ in a test with a one-sided alternative hypothesis. If 80 schools can be recruited to participate in the trial then the required number of pupils per school as calculated from Equation (4.14) is equal to $n_1 = 12.55$, which is rounded upward to $n_1 = 13$.

Alternatively, the researcher may fix the number of pupils per school and calculate the required number of schools. For instance, if 25 pupils per school are recruited, then the required number of schools as calculated from Equation (4.13) is equal to $n_2 = 64.79$, which is rounded upward to the nearest even value 66. Eventually, the number of schools can be increased with 11% to correct for varying school sizes. In that case 73.26 schools are required, which is rounded upward to 74 schools. An even number of schools is used since Equation (4.13) assumes both treatment conditions have the same number of schools. Using unequal randomization ratios makes sense when the cost ratio differs considerably from 1. For the trial by Ausems et al. (2002), the costs at the school level were 94 for a school in the control condition and 96 for a school in out-of-school intervention condition. In this case, the cost ratio is very close to 1 and the use of an equal randomization ratio is justified.

In case where neither the school size nor the number of schools is fixed, the optimal allocation of units can be calculated on basis of the budgetary constraint (4.16). The mean costs at the pupil level were 10 and the mean costs at the school level were $(94+96)/2 = 95$. The optimal number of pupils per school as calculated from Equation (4.18) is equal to $n_1^* = 7.25$. The corresponding number of schools depends on the available budget and is calculated as $C/(c_1 n_1 + c_2)$. If the budget is equal to $C = 10000$, for instance, the number of schools is equal to $n_2 = 59.7$. Both sample sizes are non-integers and have to be rounded. Rounding downward the number of pupils per school to $n_1 = 7$ allows us to sample $n_2 = 60$ schools. The costs are then equal to $C = 9900$ and the corresponding standard error of the treatment effect is 1.06. Rounding upward the number of pupils per school to $n_1 = 8$ allows us to sample $n_2 = 56$ schools. In this case the costs are $C = 9800$ and once again the corresponding standard error of the treatment effect is 1.06. Rounding upward results in slightly lower costs and is therefore the preferred design from a financial view.

5

Improving statistical power in cluster randomized trials

As was shown in the previous chapter a cluster randomized trial is less efficient in terms of estimating the treatment effect than a simple randomized trial with the same total sample size. The degree of efficiency loss increases when the cluster size and the intraclass correlation coefficient increase. One strategy to increase design efficiency is using larger sample sizes at the subject and/or cluster level. The drawback of this strategy is that it will most likely result in increasing study costs, from adding more clusters and subjects within clusters. The aim of this chapter is to discuss alternative strategies to improve statistical power in cluster randomized trials. Taking account of relevant covariates in the multilevel model is a relatively easy method. Relevant covariates are correlated with the outcome variable and they explain part of the unexplained variance. This will in general result in a lower standard error of the treatment effect estimator, provided the covariate is uncorrelated with treatment condition. For a large number of clusters, random assignment of clusters to treatment condition will most likely ensure treatment condition and the covariate will be uncorrelated. For a small number of clusters, one must rely on other randomization procedures to achieve covariate balancing such as minimization, matching and pre-stratification.

Power may also be increased by taking repeat measurements on the outcome variable. This may be done at the cluster level such that a different set of subjects within each cluster is measured at different time points, or at the subject level such that the same group of subjects within each cluster is measured at all time points. With such a design, a cluster's treatment condition remains unchanged during the course of the trial. This is not the case for cluster randomized trials with crossover where clusters are randomized to sequences of treatment conditions instead of one single treatment. Crossover may be done at the cluster level such that different sets of subjects within each cluster are assigned to each treatment, or at the subject level such that the same set of subjects within each cluster receives all treatments. Attention is also paid to the stepped-wedge design, which can be considered a special case of the crossover design. With such a design the intervention is sequentially rolled out to clusters over a number of time periods so that by the end of the trial all clusters have received the intervention.

5.1 Inclusion of covariates

Relevant covariates are variables that are supposed to have a relation to the outcome variable of interest. It may be obvious that each trial has its own set of covariates. Relevant covariates in a cluster randomized trial designed to train unemployed persons to find a job may be gender, ethnicity, educational degree, duration of unemployment and motivation. Relevant covariates in a cluster randomized trial in primary care to reduce the incidence of cardiovascular disease may be gender, age, smoking status, blood pressure and body mass index.

Allison (1995) and Allison, Allison, Faith, Paultre, and Pi-Sunyer (1997) studied the benefits and costs of covariate adjustment in simple randomized trials with clustering in neither condition. They concluded that including a covariate is a less expensive strategy than increasing sample size when the correlation between the covariate and outcome is large and the costs for measuring the covariate are low. A similar finding can be shown to hold for cluster randomized trials.

Various authors have studied the effect of the inclusion of a covariate in a cluster randomized trial (Bloom, 2005; Bloom, Richburg-Hayes, & Black, 2007; Konstantopoulos, 2012; Moerbeek, 2006b; Murray & Blitstein, 2003; Raudenbush, 1997; Raudenbush, Martinez, & Spybrook, 2007; Teerenstra, Eldridge, Graff, De Hoop, & Borm, 2012). It turns out that a higher increase in power can be achieved if the covariate has a stronger association with the outcome. Strong associations are often found when the covariate is the outcome itself, but measured at baseline. In that case, the sample size is simply the sample size of a cluster randomized trial multiplied by a factor $1 - r^2$, where

$$r = \frac{n_1\rho}{1+(n_1-1)\rho}r_c + \frac{1-\rho}{1+(n_1-1)\rho}r_s.$$

Here ρ and n_1 are as usual the intraclass correlation and the cluster size, while r_c and r_s are the cluster and subject autocorrelation (Teerenstra et al., 2012). The subject autocorrelation is the correlation of the outcome of a subject between baseline and follow-up measurement, given the cluster. The cluster autocorrelation is the correlation between the "ideal" cluster average of a cluster between baseline and follow-up, where "ideal" means without the nuisance due to sampling subjects, which is the limit of that cluster average when the number of subjects increases to infinity.

The covariate does not necessarily have to be the baseline measurement but can be any variable. According to Bloom (2005) and Raudenbush et al. (2007) associations between a covariate and the outcome tend to be much stronger between clusters than within clusters. This section therefore restricts discussion to the effects of including a cluster level covariate. This may be an aggregated variable such as mean socio-economic status within a general practice or mean pre-test scores within a school class.

The multilevel model relates outcome y_{ij} of subject i in cluster j to treatment condition x_j and covariate z_j:

$$y_{ij} = \tilde{\gamma}_0 + \tilde{\gamma}_1 x_j + \tilde{\gamma}_2 z_j + \tilde{u}_j + \tilde{e}_{ij}. \tag{5.1}$$

The regression coefficients and random terms are indicated by a tilde to distinguish them from their counterparts in a multilevel model without a covariate, as given by Equation (4.3). The variances of the random terms at the cluster and subject level are denoted $\tilde{\sigma}_u^2$ and $\tilde{\sigma}_e^2$. Treatment effect is estimated as $\hat{\tilde{\gamma}}_1 = \bar{y}_e - \bar{y}_c - \hat{\tilde{\gamma}}_2(\bar{z}_e - \bar{z}_c)$, where $\bar{y}_e$ and $\bar{y}_c$ are the mean outcomes in the experimental and control conditions and $\bar{z}_e$ and $\bar{z}_c$ are the covariate means in the experimental and control conditions. It can be verified easily that the treatment effect estimator is equal to that in a model without a covariate if the covariate means do not differ across treatment conditions. In the remainder of this section it is assumed that this is indeed the case. The standard error of the treatment effect estimator can then be shown to be equal to

$$s.e.(\hat{\tilde{\gamma}}_1) = \sqrt{\frac{4(\tilde{\sigma}_e^2 + n_1\tilde{\sigma}_u^2)}{n_1 n_2}}, \tag{5.2}$$

given -0.5 and 0.5 coding of the treatment condition x_j (Moerbeek, Van Breukelen, & Berger, 2001b). This formula resembles the formula for the standard error in a model without a covariate as given by Equation (4.8). The relation between the variance components in a model with and without the cluster level covariate is derived on basis of the method given by Snijders and Bosker (1994) and is equal to

$$\tilde{\sigma}_u^2 = (1 - \rho_B^2)\sigma_u^2 \text{ and } \tilde{\sigma}_e^2 = \sigma_e^2, \tag{5.3}$$

where ρ_B is the between-cluster residual correlation between the outcome and covariate. It is observed that a cluster level covariate explains part of the variance at the cluster level but not at the subject level. This is obvious since the value of a cluster level covariate is constant for all subjects within the same cluster. The higher the correlation ρ_B, the higher the reduction of the cluster level variance component and hence the lower the standard error (5.2). It therefore seems that a model without a covariate always results in a lower power level than a model with a covariate, but this is not always true. Recall that the degrees of freedom for the test on treatment effect are equal to $n_2 - 2$ for the model without the covariate. One additional degree of freedom is lost by the computation of the association between the outcome and covariate; thus there are $n_2 - 3$ degrees of freedom for the model with the covariate. The effect of the loss of one degree of freedom is small to negligible unless a small number of clusters is used in the trial. In such a case, including a covariate is not a blessing but a burden since it may result in an unwanted decrease of power as compared to the model without the covariate. It is therefore important to carefully select the best set of covariates based on findings in the literature and subject matter knowledge.

The effects of covariate adjustment on power to detect a small standardized effect size in a test with a two-sided alternative and $\alpha = 0.05$ are shown in Figure 5.1. The plots are based on 20 clusters, which is a rather small number and hence the t distribution is needed to calculate the power level. The results for other numbers of clusters are very similar and not shown here. Three different cluster sizes ($n_1 = 5, 30, 50$) and three different intraclass correlation coefficients ($\rho = 0.025, 0.05, 0.10$) are used. The between-cluster residual correlation ρ_B varies between 0 and 1 and is given on the horizontal axis of each subplot. Power levels for the model without a covariate are represented by the solid lines and, as they do not depend on the correlation ρ_B, they are horizontal. Power increases with decreasing intraclass correlation and increasing cluster size. This finding was also discussed in Section 4.3.1 of the previous chapter.

The power levels for the model with a covariate are represented by dashed lines, and power increases with increasing correlation ρ_B. For small values ρ_B, the inclusion of a covariate results in a lower power level, but the loss in power is very small and hardly visible in the figures. For any value of the intraclass correlation coefficient, the largest increase in power is achieved when the cluster size is largest. Similarly, for any cluster size the largest increase in power is achieved for largest intraclass correlation coefficients.

This result can also be found on basis of mathematical expressions. Substitution of Equation (5.3) into (5.2) and rearranging results in

$$s.e.(\hat{\hat{\gamma}}_1) = \sqrt{4\sigma^2 \frac{1 + (n_1 - 1)\rho(1 - \rho_B^2)}{n_1 n_2}}, \tag{5.4}$$

where σ^2 is the total variance of the outcome. The largest decrease in the standard error of the treatment effect estimator, and hence the highest increase in power, is achieved when the cluster size and intraclass correlation are largest.

Moerbeek (2006b) extended the results by including the costs for including a cluster and the costs for measuring the additional covariate. She studied two strategies to increase statistical power in cluster randomized trials: increasing the number of clusters and adding a covariate. The latter strategy is the least expensive when the cluster size and between-cluster residual correlation between the covariate and outcome are large and when the costs for adding a cluster and the costs for measuring the covariate are small. Unfortunately a general recommendation in terms of ρ_B cannot be given.

It is recommended that researchers who conduct cluster randomized trials carefully report the costs and reduction in variance components due to the inclusion of covariates to help others in designing future experiments. An overview of papers that report variance component estimates is given in Chapter 11 of this book.

A pilot study may be conducted if one cannot rely on parameter estimates as published in the literature. Such a pilot study may help to gain insight in the reduction of the variance components due to the inclusion of the covariates.

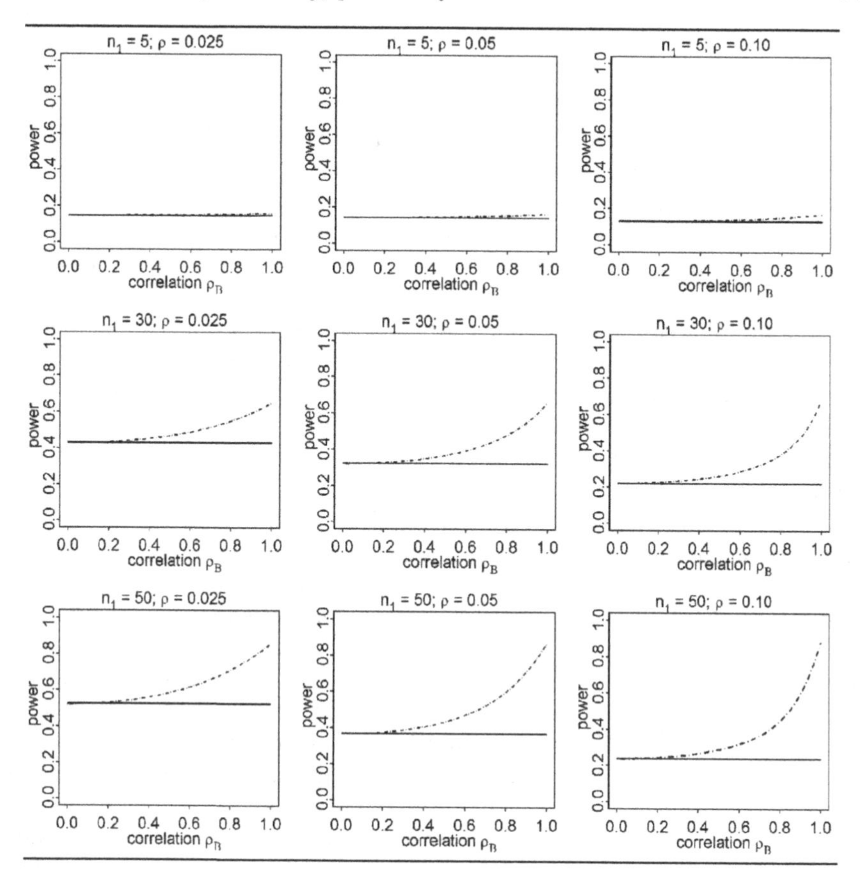

Figure 5.1: Power for the multilevel model without (solid line) and with (dashed line) a covariate.

If a particular covariate has zero or very weak correlation to the outcome one may save costs by not collecting this covariate in the remainder of the study.

5.2 Minimization, matching, pre-stratification

If there are many clusters, randomization without restrictions, that is, a completely randomized design (Donner & Klar, 2000b), is simple to implement and will with large probability result in all variables (measured and unmeasured) being equally distributed over treatment arms (i.e., covariate balance).

If the number of clusters is small, imbalances between treatment arms may more easily arise. Also, the precision (power) may be insufficient because it is largely related to the number of clusters. Posing restrictions on the randomization may therefore serve two aims: promoting balance and precision.

Having more precision increases the probability for a study to be conclusive: the relevance (or non-relevance) of an effect can be more easily established since the confidence intervals will be tighter. Having balance helps to make the trial results more credible and hence acceptable to the intended audience. Although imbalances between the treatment arms can be adjusted for in the sense that variables having unequal distribution between the treatment arms can be included as covariates in the analysis model, "the persuasiveness of the results may be reduced if the conclusions depend upon extensive statistical manipulation" (Smith & Morrow, 1991). This holds especially if a substantial difference in treatment effect is observed between the unadjusted and adjusted analysis. The question of the treatment arm likely to be favored by the observed imbalances in prognostic factors and the extent should then be discussed. This could include conducting analyses under several reasonable assumptions (sensitivity analyses) and assessing whether the results of the sensitivity analyses are generally in line with the main analysis or not.

Stratified randomization (Donner & Klar, 2000b), also called pre-stratification, or simply stratification, is one means to promote balance for a specific set of variables. For example, if location, size, and wealth are the most important cluster level prognostic factors for the outcome under consideration, the clusters are categorized according to each of these stratification factors (e.g., urban versus rural, large versus small, high versus low mean income). Subsequently all 8 ($= 2^3$) combinations of the categories of the stratification factors are formed: urban and large and low, urban and large and high, urban and small and low, urban and small and high, rural and large and low, rural and large and high, rural and small and low, and rural and small and high. These are called the (stratification) strata or cells of the design. Subsequently, randomization is performed within each stratum. Usually, this stratified randomization is combined with block randomization (not to be confused with the terminology of blocking and randomized block design as explained in Section 1.1). Then the randomization is such that, if a certain number of clusters (a 'block') is recruited into a stratum, equal numbers of clusters have been allocated to each arm within that stratum. For example, within the first four urban and large and high clusters, two will be randomized to the experimental condition and two to the control and similarly for the next four clusters, etc. While stratified block randomization is good at achieving balance, it will only increase precision (power) if the gain of homogeneity within a stratum versus the heterogeneity over the several strata is large. This requires that there is a strong correlation between the stratification factors and the outcome (Lewsey, 2004). Because the number of strata increases rapidly with the number of stratification factors (2, 4, 8, 16 strata for 1, 2, 3, 4 stratification factors), relatively few stratification factors can be included. S. Eldridge, Cryer, Feder, and Underwood (2001) provide a practical example.

Matched-pair randomization is a special case of the stratified randomization, where each stratum consists of only two clusters (a pair). Given a set of available clusters, it may be conceptually easier for researchers or trialists to think which pairs of clusters match than to stratify according to several factors, for instance, because some strata will be sparse or even empty in stratified randomization. However, matched-pair randomization may be less efficient than it seems when the number of clusters is small. First, each stratum counts as a loss of approximately 1 degree of freedom in the analysis in stratified designs. For matched-pair randomization, this means that the loss of degrees of freedom is roughly half the number of clusters, which means that the threshold for statistical significance (the critical value of the statistical test) is easily higher than in an completely randomized design when the number of clusters is small. Therefore, the gain in homogeneity within a pair (correlation within a pair) has to be even larger than the homogeneity within a stratum in stratified randomization for matching to be worthwhile (Martin, Diehr, Perrin, & Koepsell, 1993).

In fact, it may sometimes pay off to match in the randomization but break the matches in the analysis (Diehr, Martin, Koepsell, & Cheadle, 1995). Besides this, there are also restrictions in the analysis of matched-pair design (Klar & Donner, 1997). If good baseline data of the clusters are available, they may be used to make an informed decision which design is more powerful. Hughes (2005) proposed a method for choosing between a completely randomized and a matched-pair design. Sample size formulae for the completely randomized design and the matched-pair design are provided by Hayes and Bennett (1999). Note that these authors parameterize the degree of clustering not with the intraclass correlation but with the coefficient of variation $cv = sd/m$, that is, they relate the *standard deviation* sd of the *true* cluster rates (or proportions or means) to the average m of these rates (proportions or means) instead of relating the *variance* of these cluster rates (or proportions or means) to the total variance of the outcome, like the intraclass correlation does. Especially for types of outcomes that have a relation between the variance and the mean such as the Poisson distribution for rates (variance = mean) and the binomial distribution for proportions (variance = mean times 1 minus mean), this parameterization simplifies formulae.

Another means to promote balance for a specific set of variables is minimization. In contrast with stratified randomization that promotes balance within each combination of stratification factors (i.e., within each stratum), minimization usually only promotes balance between treatment arms overall (marginal balance). To illustrate this, revisit the example above. Stratified randomization aims to ensure that from the urban and large and low income clusters, half will be allocated to the experimental and half to the control intervention. The same holds for urban and large and high income clusters, and the other strata. Therefore, there is balance within strata. In successful minimization, half of the urban clusters will be in the experimental condition, half of the rural clusters will be in the experimental condition, and the same for the large clusters, the small clusters, the low income clusters and the high income clusters. Thus, there is balance within each level of each stratification factor (location, size, and wealth) separately, that is, marginal balance.

The advantage of minimization over stratification is that it allows a researcher to include more minimization factors. Common to all minimization methods is that a measure of imbalance between treatment arms is calculated and allocation aims to minimize this imbalance. Minimization methods differ in which measure of imbalance they use and how they allocate to minimize this imbalance measure (Pocock & Simon, 1975; Scott, McPherson, Ramsay, & Campbell, 2002). For example, minimization can be done sequentially: every time a new cluster is included, the imbalance arising when that cluster is allocated to the experimental arm is calculated, and also the imbalance for assigning that cluster to the control arm is calculated. Then in deterministic minimization, the cluster is assigned to the arm with the smallest of those two imbalances (or if the imbalance measure is equal in both arms: with equal probability to either of the arms). In minimization with a random element, this allocation is done with high probability (e.g., 3/4) to the arm with the smallest of the two calculated imbalances.

Especially when the number of clusters is small, it is often possible to know the clusters and hence their characteristics in advance (i.e., before starting the trial). In that case, an even more optimal method of balancing can be employed. For every possible allocation of the clusters to the treatment arms, the imbalance between arms is calculated. Subsequently, one allocation is randomly chosen from the set of allocations having the smallest imbalance (De Hoop, Teerenstra, Van Gaal, Moerbeek, & Borm, 2012; Carter & Hood, 2008; Nietert, Jenkins, Nemeth, & Ornstein, 2009). If this set contains only a few allocations, then it can be enlarged by considering the allocations having the top 10% smallest imbalances (10th percentile). This compares favorably to unrestricted randomization, sequential minimization, and matching in terms of balance achieved (De Hoop et al., 2012). As for the analysis, most tests (e.g., t tests, ANOVA and Wald tests in regression models) are theoretically based on an unrestricted randomization. As minimization restricts the randomization, the type I error of such classical tests could be increased above nominal levels. Therefore, one should investigate that sufficient randomness remains in the design (Moulton, 2004). As an alternative, a re-randomization test could be used, since it is expected to keep the type I error at nominal levels.

5.3 Taking repeated measurements

The multilevel model for cluster randomized trials presented in the previous chapter is appropriate for the so-called posttest-only design. With this design one measurement on the outcome variable at posttest is taken and related to treatment condition in a regression model. This design can be extended by taking repeated measurements across time. Repeat observations may be taken at the cluster level or subject level. Taking repeat measurements at the cluster level implies that a different set of subjects within each cluster is measured at each time point. Such a design is referred to as a cross-sectional design. On the other hand, the same set of subjects within each cluster is measured when repeated measures are taken at the subject level and this is called a cohort design. For both designs, the statistical model is presented and it is shown under which conditions taking repeat measurements results in an increase of statistical power.

This section is based on the book (Murray, 1998) and papers (Murray, Hannan, Wolfinger, Baker, & Dwyer, 1998; Murray, 2001; Murray & Blitstein, 2003; Murray, Van Horn, Hawkins, & Arthur, 2006) by Murray and coworkers, who have extensively studied the benefits of taking repeat measurements. The models presented in this section are based on the multilevel modeling framework. For the corresponding analysis of variance models, the reader is referred to the book and papers by Murray. As in the work by Murray, a restriction is made to two measurements: before and after the intervention is delivered. This design is called a pretest-posttest design, irrespective of whether repeat measurements are taken at the cluster or subject level.

The model for the pretest-only design is given by Equation (4.3) with corresponding standard error of the treatment effect estimator (4.8). The treatment effect is calculated as the difference in mean outcomes in the experimental and control conditions and the test on treatment effect follows a t distribution with $n_2 - 2$ degrees of freedom under the null hypothesis of no treatment effect.

For the cross-sectional design the model for subject i in cluster j is given by

$$y_{ij} = \tilde{\gamma}_0 + \tilde{\gamma}_1 x_j + \tilde{\gamma}_2 t_{ij} + \tilde{\gamma}_3 x_j t_{ij} + \tilde{u}_{0j} + t_{ij}\tilde{u}_{1j} + \tilde{e}_{ij}, \quad (5.5)$$

where t_{ij} is the time indicator that varies at the subject level and x_j is the treatment condition that varies at the cluster level. As previously, treatment condition is coded -0.5 for the control group and 0.5 for the intervention group.

This model is an extension of the model (4.3) for the posttest-only design since it includes a fixed effect for time t_{ij}, a fixed effect for the time by treatment interaction $t_{ij}x_j$ and a random interaction $t_{ij}\tilde{u}_{1j}$ of time and cluster. The regression coefficients and random effects are indicated with a tilde to distinguish them from their counterparts in the model (4.3) without the pretest measurement. The random effects are assumed to follow a normal distribution: $\tilde{e}_{ij} \sim N(0, \tilde{\sigma}_e^2)$, $\tilde{u}_{0j} \sim N(0, \tilde{\sigma}_{u0}^2)$ and $\tilde{u}_{1ij} \sim N(0, \tilde{\sigma}_{u1}^2)$. The variance at the subject level is equal to $\tilde{\sigma}_e^2$ whereas the variance at the cluster level consists of two components $\tilde{\sigma}_{u0}^2$ and $\tilde{\sigma}_{u1}^2$. The over-time correlation within clusters is calculated as $\rho_{cluster} = \tilde{\sigma}_{u0}^2/(\tilde{\sigma}_{u0}^2 + \tilde{\sigma}_{u1}^2)$ and is non-negative.

The unstandardized effect of the intervention is calculated as the difference in change scores $(\bar{y}_{e2} - \bar{y}_{e1}) - (\bar{y}_{c2} - \bar{y}_{c1})$, where subscripts e and c are used to denote the experimental and control conditions and subscripts 1 and 2 are used to indicate the pretest and posttest. The treatment effect estimate is the same as that for the posttest-only design if the mean pretest scores are equal across the two treatments. In the remainder of this section, it is assumed that this is indeed the case.

The standard error of the treatment effect estimator can be expressed in terms of the variance components for the posttest-only design:

$$s.e.(\hat{\gamma}_1) = \sqrt{2\frac{4(\sigma_e^2 + n_1\sigma_u^2(1-\rho_{cluster}))}{n_1 n_2}}. \tag{5.6}$$

A comparison with the corresponding Equation (4.8) for the posttest-only design reveals an extra multiplication factor 2 in the argument of the square root. This extra 2 is included since the intervention effect estimator is based on twice as many time points as in a posttest-only design. Furthermore it is observed that the variance component at the cluster level is multiplied by $(1-\rho_{cluster})$. Since the degrees of freedom of the test statistics under the null hypothesis of no treatment effect are equal to those of a posttest-only design (i.e., $n_2 - 2$) the power of the cross-sectional design is larger than the power of the posttest-only design when the reduction $(1-\rho_{cluster})$ in the variance component at the cluster level is large enough to make up the extra 2. It is shown under which conditions this is actually the case after the formulation of the statistical model for the cohort design.

With the cohort version of the pretest-posttest design, two repeated measurements are taken on each subject: one at pretest and one at posttest. The multilevel model therefore requires an additional level of nesting, namely the repeated measures level, which is located below the subject level. The model for measurement h in subject i in cluster j relates the outcome variable y_{hij} to time point treatment x_j and t_{hij}:

$$y_{hij} = \tilde{\tilde{\gamma}}_0 + \tilde{\tilde{\gamma}}_1 x_j + \tilde{\tilde{\gamma}}_2 t_{hij} + \tilde{\tilde{\gamma}}_3 x_j t_{hij} + \tilde{\tilde{u}}_{0j} + t_{hij}\tilde{\tilde{u}}_{1j} + \tilde{\tilde{e}}_{0ij} + t_{hij}\tilde{\tilde{e}}_{1ij} + \tilde{\tilde{m}}_{hij}. \tag{5.7}$$

The regression weights and random effects are indicated with a double tilde to distinguish them from their counterparts in the posttest-only design and the cross-sectional version of the pretest-posttest design. Treatment condition is a variable that varies at the cluster level; time is a variable that varies at the repeated measures level. The effect of time is allowed to vary randomly over subjects and clusters; thus the model includes a random slope at the measurement and subject level.

The multilevel model contains main and interaction effects of treatment and time in the fixed part. The random part consists of main effects at the measurement ($\tilde{\tilde{m}}_{hij}$), subject ($\tilde{\tilde{e}}_{0ij}$) and cluster level ($\tilde{\tilde{u}}_{0j}$) and of interactions between time and subject ($t_{hij}\tilde{\tilde{e}}_{1ij}$) and time and cluster ($t_{hij}\tilde{\tilde{u}}_{1j}$). It is assumed $\tilde{\tilde{m}}_{hij} \sim N(0, \tilde{\tilde{\sigma}}_m^2)$, $\tilde{\tilde{e}}_{0ij} \sim N(0, \tilde{\tilde{\sigma}}_{e0}^2)$, $\tilde{\tilde{u}}_{0j} \sim N(0, \tilde{\tilde{\sigma}}_{u0}^2)$, $\tilde{\tilde{e}}_{1ij} \sim N(0, \tilde{\tilde{\sigma}}_{e1}^2)$ and $\tilde{\tilde{u}}_{1j} \sim N(0, \tilde{\tilde{\sigma}}_{u1}^2)$. The over-time correlations within clusters and within subjects are calculated as $\rho_{cluster} = \tilde{\tilde{\sigma}}_{u0}^2/(\tilde{\tilde{\sigma}}_{u0}^2 + \tilde{\tilde{\sigma}}_{u1}^2)$ and

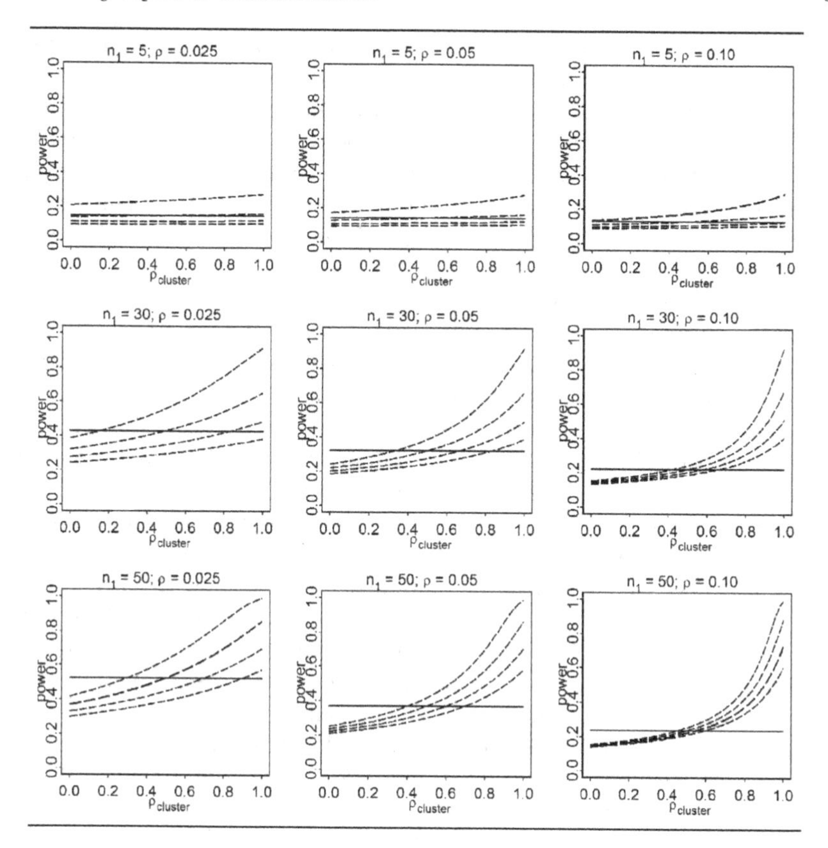

Figure 5.2: Power for posttest-only design (horizontal solid line) and the pretest-posttest design (curved dashed lines). From bottom to top the dashed lines represent power levels for $\rho_{subject} = 0, 0.25, 0.5$, and 0.75.

$\rho_{subject} = \tilde{\tilde{\sigma}}^2_{e0}/(\tilde{\tilde{\sigma}}^2_{e0} + \tilde{\tilde{\sigma}}^2_{e1} + \tilde{\tilde{\sigma}}^2_{m})$. By definition, these correlations are non-negative.

The treatment effect is estimated as for the cross-sectional design and the corresponding standard error is equal to

$$s.e.(\hat{\gamma}_1) = \sqrt{2\frac{4(\sigma^2_e(1-\rho_{subject}) + n_1\sigma^2_u(1-\rho_{cluster}))}{n_1 n_2}}. \tag{5.8}$$

This standard error is smaller than the standard error for the cross-sectional version of the pretest-posttest design since it shows a reduction in the cluster level variance component and also in the subject level variance component. If these reductions are large enough, the cohort design will result in a higher power than the posttest-only design.

Figure 5.2 shows the power to detect a small standardized effect for the posttest-only design (solid lines), the cross-sectional design (lowest dashed line within each plot) and the cohort design (other dashed lines). The power calculations are based on a total of $n_2 = 20$ clusters in a test with a two-sided alternative and $\alpha = 0.05$. Cluster size varies across the rows of this figure and intraclass correlation coefficient across the columns. The same values for model parameters and sample sizes were used in Section 5.1. The horizontal axis within each plot represents the over-time correlation at the cluster level; the dashed lines represent different values of the over-time correlation at the subject level.

The power for the posttest-only design does not depend on the over-time correlations and has the same value as in the model without covariates in the previous section (see Figure 5.1). This power increases with increasing cluster size and decreasing intraclass correlation coefficient.

Taking repeat measurements does not always guarantee a higher power level is achieved. Only if the over-time correlations at the cluster and subject level are sufficiently large can a larger power level be achieved with the pretest-posttest design than with the posttest-only design. For any intraclass correlation coefficient, the highest increase in power is achieved when cluster size is large. Taking repeat measurements in a trial with clusters sizes as small as 5 hardly results in an increase in power. For large values of the intraclass correlation it can be stated that the pretest-posttest design is more powerful than the posttest-only design when the over-time correlation at the cluster level is about 0.5 or higher. Only for such large cluster level over-time correlations does the difference in power levels for different values of the over-time correlation at the subject level becomes explicit. For smaller values of the intraclass correlations the minimal required cluster level over-time correlation depends on the size of the over-time correlation at the subject level. For low values of the over-time correlation at the cluster level, a higher over-time correlation at the subject level is needed, and vice versa.

5.4 Crossover in cluster randomized trials

In a crossover trial subjects are assigned to sequences of treatments rather that one single treatment to study the effect of treatment within rather than between subjects. The most common crossover design is the AB/BA design where one group of subjects receives treatment A in period 1 and treatment B in period 2 and the other group of subjects receives the two treatments in reverse order. Crossover designs are in particular useful for medical trials that aim to compare various treatments for chronic disease. The aim of treatment is not to cure the disease but to suppress its symptoms.

Consider as an example a pilot study that aimed to test whether dexmethylphenidate (dMPH) could facilitate tic suppression in children and adolescents with attention-deficit/hyperactivity disorder and Tourette's Disorder (Lyon et al., 2010). Trial subjects were given dMPH on one visit and no medication on another, using a crossover design, and on both days tic suppression was measured.

Examples can also be found in the social sciences. Widenhorn-Muller, Hille, Klenk, and Weiland (2008) studied the effect of breakfast on the cognitive performance and mood in high school students. One group of students received breakfast on the morning of the first testing day and no breakfast on the morning of the second testing day, which was 7 days later, while the reverse order was used in the second group of students. Single or double blinding may be used in crossover trials. In the first example trial it may probably be possible to keep both the trial subjects and the observers unaware of the treatments, but in the second example trial it is impossible to blind subjects.

The crossover design is often compared with the parallel group design, where just one treatment is available within each subject. The main advantage of the crossover design over the parallel group design is that treatment conditions are compared within rather than between subjects. As a result the between-subject variability is eliminated from the variance of the treatment effect estimator, resulting in higher power of the test on treatment effect. The main disadvantage of the crossover design is that it may be hampered by carry-over effects. Carry-over occurs when the effect of a treatment given in a certain period lasts into the subsequent period(s). Carry-over may result in an interaction of period and treatment conditions and this may lead to incorrect conclusions about treatment effects. One way to prevent carry-over is to implement a wash-out period of sufficient length between periods. In the remainder of this section it is assumed that carry-over effects are absent.

The statistical model for the AB/BA crossover design where the possible nesting of subjects within clusters is ignored is introduced first and it is shown that this design is more efficient than the parallel group design. A more extensive introduction to crossover trials can be found in Jones and Kenward (2002) and Senn (2002). The design of crossover trials is extensively discussed in Chapter 8 of Berger and Wong (2009).

It is assumed a single measurement is taken at the end of the first and second periods. The two treatment conditions are referred to as the experimental and control conditions. The multilevel model for measurement $h = 1, 2$ in subject i is given by

$$y_{hi} = \gamma_0 + \gamma_1 x_{1hi} + \gamma_2 x_{2h} + e_i + m_{hi}. \tag{5.9}$$

As usual, y_{hi} is the (continuous) outcome variable. The predictor x_{1hi} denotes the treatment condition of subject i in the h-th period and is coded -0.5 and 0.5 for the control and experimental groups, respectively. Thus, $x_{11i} = -0.5$ and $x_{12i} = 0.5$ for subjects who receive the control in the first period and the experimental condition in the second, and $x_{11i} = 0.5$ and $x_{12i} = -0.5$ for subjects who receive the treatments in reverse order. The predictor x_{2h} denotes the measurement or period and is equal to 0 for the first period and 1 for the second. The regression coefficients γ_1 and γ_2 represent the effects of treatment and period and γ_0 is the intercept that represents the mean outcome in the first period. It should be noted that an interaction between treatment and period is not included in the model. In other words, the effect of treatment is assumed to be constant across periods.

The random effects e_i and m_{hi} represent random error at the subject and measurement level. They are assumed to be independent of each other and to follow a normal distribution with zero mean and variances σ_e^2 and σ_m^2. They sum up to the total error variance $\sigma^2 = \sigma_e^2 + \sigma_m^2$. The intra-subject correlation $\rho_1 = \sigma_e^2/(\sigma_e^2 + \sigma_m^2)$ calculates the proportion of the total error variance accounted for by the between-subject variation.

For each subject, the effect of treatment can be estimated by the difference in both outcome scores; the mean difference across all subjects is the overall treatment effect. This estimator has standard error $s.e.(\hat{\gamma}_1) = \sqrt{2\sigma_m^2/n} = \sqrt{2\sigma^2(1-\rho_1)/n}$, where n is the total number of subjects (B. W. Brown, 1980). This formula holds when equal numbers of subjects are assigned to treatment sequences AB and BA. The treatment effect estimator of a parallel group design, with one treatment per subject, is larger: $s.e.(\hat{\gamma}_1) = \sqrt{4\sigma^2/n}$. The efficiency of a crossover design relative to a parallel group design with the same number of subjects is defined as the ratio of the squared standard errors (i.e., variances) of $\hat{\gamma}_1$:

$$RE = \frac{1}{2}(1-\rho_1), \tag{5.10}$$

and varies between $\frac{1}{2}$ and 0. In other words, at least twice as many subjects are needed in a parallel group design to be as efficient with respect to estimating the treatment effect as the crossover design. Crossover can therefore be considered to be a powerful tool to increase efficiency in simple randomized trials.

The null hypothesis $H_0 : \gamma_1 = 0$ is tested with the test statistic $t = \hat{\gamma}_1/s.e.(\hat{\gamma}_1)$ which follows a central t distribution with $n-2$ degrees of freedom if the null hypothesis is true. Under the alternative hypothesis it follows a non-central t distribution with non-centrality parameter

$$\lambda = \frac{\gamma_1}{\sqrt{\frac{2\sigma_m^2}{n}}}. \tag{5.11}$$

The total number of subjects n to achieve a power level $1-\beta$ to detect a treatment effect of size γ_1 in a test with a two-sided alternative and type I error rate α can be approximated by using the standard normal distribution

$$n = \frac{2\sigma_m^2(z_{1-\alpha/2} + z_{1-\beta})^2}{\gamma_1^2}. \tag{5.12}$$

For a given total sample size n unequal allocation of subjects to treatment sequences results in the same power level. If the sample size in the one sequence is fixed beforehand, it can be subtracted from the required n to calculate the required number of subjects in the other condition.

If crossover is used in cluster randomized trials a decision has to be made if crossover is done at the cluster or individual level. In the first case different sets of subjects are used in the two periods, while in the second case the same set of subjects is used in both periods. The first design is referred to as a cluster randomized cluster crossover design and the second design as a cluster randomized individual crossover design. Harrison and Brady (2004) and Giraudeau, Ravaud, and Donner (2008) studied crossover at the cluster level; an extension to crossover at the individual level was made by Rietbergen and Moerbeek (2011). The formulae that follow hold for trials where as many clusters are assigned to treatment sequence AB as to sequence BA.

The multilevel model for the cluster randomized cluster crossover design is given by

$$y_{hij} = \gamma_0 + \gamma_1 x_{1hi} + \gamma_2 x_{2h} + u_j + e_{ij}. \tag{5.13}$$

The predictors x_{1hi} and x_{2h} and the regression coefficients γ_0, γ_1 and γ_2 are defined as in model (5.9). Random error at the subject and cluster level is represented by the random effects u_j and e_{ij}. These effects are assumed to be independent of each other and to follow a normal distribution with zero mean and variance σ_u^2 and σ_e^2. As only one measurement per subject is taken, the model does not include a random effect at the measurement level. The total error variance of the outcome y_{ij} is calculated as $\sigma^2 = \sigma_u^2 + \sigma_e^2$. The intracluster correlation coefficient ρ_2 is the proportion of the total variation at the cluster level and is calculated as $\rho_2 = \sigma_u^2/(\sigma_u^2 + \sigma_e^2)$. It is the correlation between two randomly selected subjects in the same cluster, irrespective of whether these subjects are in the same or different periods.

For this model, the standard error of the treatment effect estimator is given by

$$s.e.(\hat{\gamma}_1) = \sqrt{4\sigma^2 \frac{1-\rho_2}{n_1 n_2}}, \tag{5.14}$$

where n_1 and n_2 are the cluster size and total number of clusters. The efficiency of a cluster randomized cluster crossover trial relative to a cluster randomized trial without crossover but with the same number of subjects can be shown to be

$$RE = \frac{1-\rho_2}{1+(n_1-1)\rho_2}. \tag{5.15}$$

The relative efficiency depends on the cluster size and varies between 0 and 1, implying that a gain in efficiency can be achieved when crossover at the cluster level is applied in a cluster randomized trial.

Giraudeau et al. (2008) studied the perhaps more realistic situation where the correlation between two randomly selected subjects in the same cluster and period is different from the correlation between two subjects in different periods (denoted by η). In this situation, the standard error of the treatment

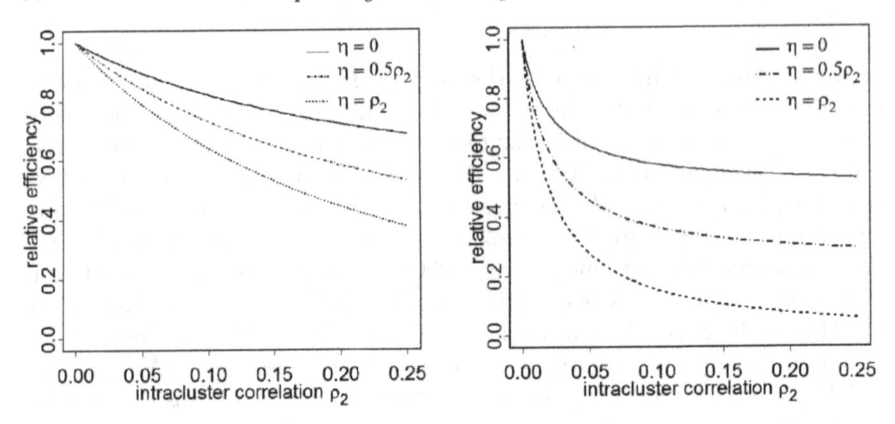

Figure 5.3: Efficiency of the cluster randomized cluster crossover design relative to the cluster randomized design as a function of ρ_2 and for different values of η. Plotted in the left figure for $n_1 = 5$ and in the right figure for $n_1 = 50$.

effect estimator is equal to

$$s.e.(\hat{\gamma}_1) = \sqrt{4\sigma^2 \frac{1 + (\frac{1}{2}n_1 - 1)\rho_2 - \frac{1}{2}n_1\eta}{n_1 n_2}}. \tag{5.16}$$

It can be easily verified that this formula reduces to Formula (5.14) in case $\eta = \rho_2$.

The efficiency relative to a cluster randomized trial with the same number of subjects is now equal to

$$RE = \frac{1 + (\frac{1}{2}n_1 - 1)\rho_2 - \frac{1}{2}n_1\eta}{1 + (n_1 - 1)\rho_2}. \tag{5.17}$$

Figure 5.3 shows the efficiency of the cluster randomized cluster crossover design relative to the cluster randomized design as a function of the intracluster correlation ρ_2 and for different values of η. A restriction is made to values $\rho_2 < 0.25$ since higher values are rather uncommon in cluster randomized trials. A small cluster size $n_1 = 5$ is used in the left plot and a higher cluster size of $n_1 = 50$ in the right plot. At $\rho_2 = 0$, the two designs are equally efficient. The efficiency of the cluster randomized cluster crossover design increases as the intracluster correlation increases, but this effect is tempered by a smaller correlation between two subjects from the same cluster but from different periods. Crossover is especially useful in trials with larger cluster sizes.

The test statistic $t = \hat{\gamma}_1 / s.e.(\hat{\gamma}_1)$ follows a t distribution with $n_2 - 2$ degrees of freedom if the null hypothesis of no treatment effect is true. For a large number of clusters, the standard normal distribution is used to calculate the required sample size. For a fixed cluster size n_1, the required number of clusters is

$$n_2 = 4\frac{1+(\frac{1}{2}n_1-1)\rho_2-\frac{1}{2}n_1\eta}{n_1}\frac{(z_{1-\alpha/2}+z_{1-\beta})^2}{\delta^2}. \quad (5.18)$$

For a fixed number of clusters n_2 the required cluster size is calculated as

$$n_1 = \frac{4(1-\rho_2)}{n_2(\frac{\delta}{z_{1-\alpha/2}+z_{1-\beta}})^2-4(\frac{1}{2}\rho_2-\frac{1}{2}\eta)}. \quad (5.19)$$

In both equations, $\delta = \gamma_1/\sigma$ is the standardized treatment effect Note that in both equations a two-sided alternative hypothesis is used and $z_{1-\alpha/2}$ should be replaced by $z_{1-\alpha}$ in case of a one-sided alternative hypothesis.

In case neither sample size is fixed beforehand, a budgetary constraint can be used to calculate the optimal allocation of units. The costs at the cluster level are c_2 which consist of implementing treatment A in the one period and treatment B in the other. The costs to measure a subject are c_1 per period. The cluster size is n_1 but each subject is measured just once. The costs are then $c_1n_1n_2 + c_2n_2$ and given a total budget C the optimal cluster size is

$$n_1 = \sqrt{\frac{c_2}{c_1}\frac{1-\rho_2}{\frac{1}{2}\rho_2-\frac{1}{2}\eta}}, \quad (5.20)$$

and the optimal number of clusters is

$$n_2 = \frac{C}{\sqrt{\frac{c_1c_2(1-\rho_2)}{\frac{1}{2}\rho_2-\frac{1}{2}\eta}}+c_2}. \quad (5.21)$$

As we will now show, efficiency can be further improved when crossover is at the individual rather than at the cluster level. With such a design, each subject is included in both periods and two measurements are taken on each subject: one at the end of the first period and another at the end of the second period. For this reason, the multilevel model includes a random effect at the measurement level. The model includes fixed effects for treatment and period and random effects at the measurement (m_{hij}), subject (e_{ij}) and cluster (u_j) level:

$$y_{hij} = \gamma_0 + \gamma_1x_{1hi} + \gamma_2x_{2h} + u_j + e_{ij} + m_{hij}. \quad (5.22)$$

The predictor variables x_{1hi} and x_{2h} and their corresponding effects are defined as for the cluster randomized cluster crossover design. The random effects are assumed to be independent of each other and to follow a normal distribution with zero mean and variances σ^2_m, σ^2_e and σ^2_u. These variances sum up to the total variance: $\sigma^2 = \sigma^2_m + \sigma^2_e + \sigma^2_u$. An intracluster correlation coefficient $\rho_2 = \sigma^2_u/(\sigma^2_m + \sigma^2_e + \sigma^2_u)$ and intrasubject correlation coefficient $\rho_1 = \sigma^2_e/(\sigma^2_m + \sigma^2_e + \sigma^2_u)$ are now distinguished.

The standard error of the treatment effect estimator of a cluster randomized individual crossover design is equal to

$$s.e.(\hat{\gamma}_1) = \sqrt{2\sigma^2 \frac{1 - \rho_1 - \rho_2}{n_1 n_2}}. \tag{5.23}$$

In other words, only the error variance at the measurement levels remains in the calculation of the variance. It should be noted that a 2 rather than a 4 appears in the numerator of the variance since it is based on two measurements per subject.

The efficiency of the cluster randomized individual crossover design relative to the cluster randomized design with the same number of subjects is equal to

$$RE = \frac{1}{2} \frac{1 - \rho_1 - \rho_2}{1 + (n_1 - 1)\rho_2}, \tag{5.24}$$

and its value lies between 0 and 0.5. Figure 5.4 shows the relative efficiency as a function of the intracluster correlation and for different values of the intrasubject correlation. A small cluster size of $n_1 = 5$ is used in the left plot; a higher cluster size of $n_1 = 50$ in the right. The cluster randomized individual crossover design is twice as efficient as the cluster randomized design when both ρ_1 and ρ_2 are equal to zero. This is explained by the fact that subjects are measured twice with the cluster randomized individual crossover design and only once with the cluster randomized design. The relative efficiency of the cluster randomized individual crossover design further increases when ρ_1 and/or ρ_2 increase and this effect is stronger for larger cluster size. It may also be mentioned that when $\rho_2 = 0$, clustering of subjects within clusters is absent and the cluster randomized individual crossover design is as efficient as the crossover design.

Again the test statistic $t = \hat{\gamma}_1 / s.e.(\hat{\gamma}_1)$ is used to test the null hypothesis of no treatment effect. If the null hypothesis is true, this test statistic follows a t distribution with $n_2 - 1$ degrees of freedom but the standard normal approximation to derive the required sample size is used. If the cluster size n_1 is fixed, the required number of clusters follows from

$$n_2 = 2\frac{1 - \rho_1 - \rho_2}{n_1} \frac{(z_{1-\alpha/2} + z_{1-\beta})^2}{\delta^2}. \tag{5.25}$$

Given a fixed number of clusters n_2 the required cluster size is calculated from

$$n_1 = 2\frac{1 - \rho_1 - \rho_2}{n_2} \frac{(z_{1-\alpha/2} + z_{1-\beta})^2}{\delta^2}. \tag{5.26}$$

In both equations, $\delta = \gamma_1/\sigma$ is the standardized effect of treatment. Again, these equations are based on a two-sided alternative hypothesis and $z_{1-\alpha/2}$ should be replaced by $z_{1-\alpha}$ in case of a one-sided alternative.

The cost function for this design is $2c_1 n_1 n_2 + c_2 n_2 = C$. Note that the factor 2 appears in the first term at the left since each subject is measured

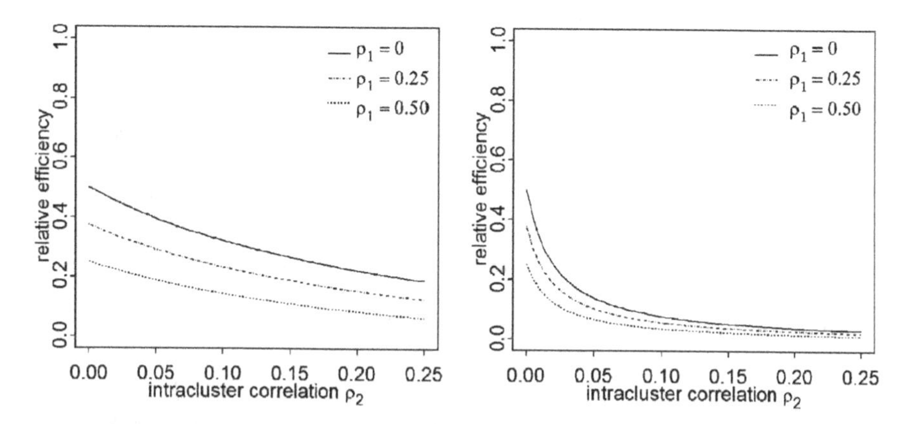

Figure 5.4: Efficiency of the cluster randomized individual crossover design relative to the cluster randomized design as a function of ρ_2 and for different values of ρ_1. Plotted in the left figure for $n_1 = 5$ and in the right figure for $n_1 = 50$.

twice. The optimal design includes just one cluster and as many subjects per cluster as allowed by the budget. As one cluster cannot be assumed to represent the population of clusters from which it is drawn, a larger number of clusters is advised. According to the rule of thumb by Snijders and Bosker (2012) $n_2 = 20$ clusters are included and the cluster size is then $n_1 = \frac{C-20c_2}{40c_1}$.

5.5 Stepped wedge designs

In a cluster randomized stepped wedge design, clusters switch from the control to the experimental condition in steps and the timing of switching is randomized over the clusters. Consider for example the Act In case of Depression (AiD) trial (Leontjevas et al., 2013) which investigated the effectiveness of a multi-disciplinary program to detect and treat depression in elderly residents (subjects) dwelling in nursing home units (clusters). At baseline (time point 0), all clusters were measured in the usual care condition (see Figure 5.5). Then the AiD program was implemented in the clusters randomized to time point 1, (group 1). At time point 1, these clusters were measured in the AiD condition while the remaining clusters were measured in the usual care condition. After time point 1, the AiD program was introduced in the clusters randomized to time point 2 (group 2). Then, at time point 2, the effect of the AiD program was measured in the clusters that had already switched by that time (groups 1 and 2), while the usual care condition was measured in

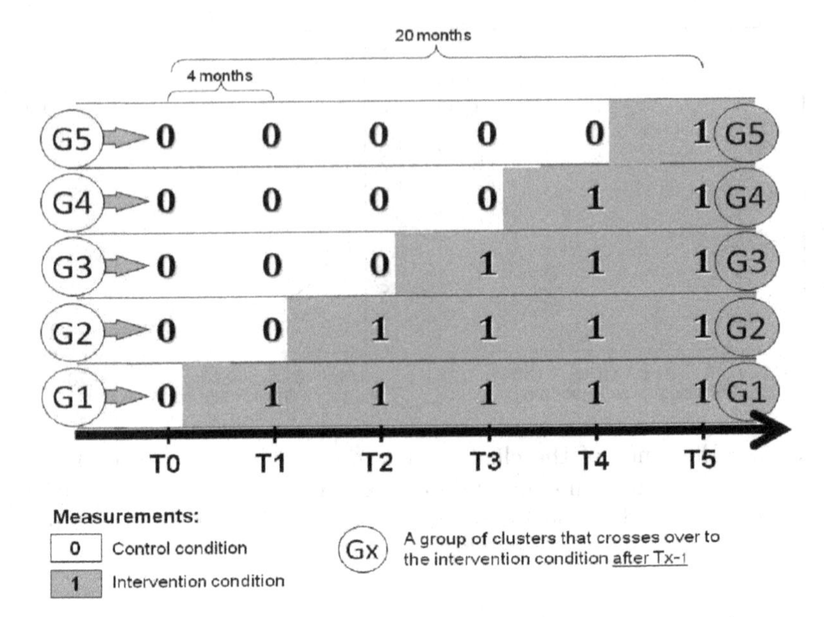

Figure 5.5: Graphical representation of a stepped wedge design. (Reprinted with permission from Leontjevas et al. (2013). © Elsevier).

the others. Proceeding this way, all the clusters successively switch to the AiD program until by time point 5 all clusters switched and were measured in the experimental condition at time 5.

There may be several reasons to opt for a stepped wedge design (Hussey & Hughes, 2007; C. A. Brown & Lilford, 2006; Mdege, Man, Taylor nee Brown, & Torgerson, 2011). First of all, a crossover trial may not be feasible if the switch is (practically) irreversible in reasonable time. For instance, the AiD program consisted of training of and collaboration between the nursing staff, recreational therapists, physician, and psychologist attached to the unit and this change in care organization can hardly be undone. Moreover, the AiD program was built using elements and insights that had either a strong rationale or had already shown effectiveness.

This illustrates a second reason to choose for a stepped wedge design: the experimental condition is strongly believed to be effective or has already shown to be effective in other, for instance more controlled, research settings and the purpose of the trial is to estimate the size of its effect in the setting of interest. Due to such belief or proof of concept, it would be unethical to withhold or withdraw the experimental condition from part of the clusters. Of course, the experimental condition could also be implemented in a parallel group design, namely after the trial period. However, this raises a third reason for a stepped wedge design: in a stepped wedge trial, clusters have the opportunity to receive the experimental condition earlier, which may be an incentive for recruitment.

Moreover, there is an another advantage here: the implementation will be *within* the trial period, which promotes the quality of the implementation, because the resources and the motivation to implement are still there (and not after the trial has finished). Also, in the AiD trial it would have been impractical to implement the AiD program in all intervention clusters from the start because of the substantial training effort required. This is a logistic advantage of the stepped wedge design over a parallel group design: the experimental condition is implemented in steps, which may be more feasible than implementing it in half of the clusters at once.

Finally, the number of clusters available in the AiD trial was limited (16 clusters), and a parallel group design with this number of clusters would not have provided sufficient power. A stepped wedge design with five steps required a more than six times smaller sample size at each step than a parallel group design with one measurement, and hence six times fewer of clusters (Woertman et al., 2013). This made the AiD trial possible.

Measurement m $(= 0, \ldots, T)$ of subject i $(= 1, \ldots, n_1)$ in cluster j $(= 1, \ldots, n_2)$ is modeled as

$$y_{mij} = \gamma_{000} + \gamma_{001} x_{jm} + \gamma_{m00} t_m + u_{0j} + t_m u_{1j} + e_{0ij} + t_m e_{1ij} \tag{5.27}$$

where $x_{jm} (= 0, 1)$ denotes the condition (0 control, 1 experimental), the cluster is in at time point m; γ_{001} is the treatment effect (averaged over the time points and clusters); γ_{000} is the grand mean, that is, the mean in control group at baseline $(m = 0)$; t_m the indicator of time point $m (= 0, \ldots, T)$ so that γ_{m00} is the change from baseline to time point m in the control group; $u_{0j} \sim N(0, \sigma^2_{u0})$ describes the variation between clusters at a given time point; $t_m u_{1j} \sim N(0, \sigma^2_{u1})$ (interaction of cluster with time) describes the variation over time of a given cluster; $e_{0ij} \sim N(0, \sigma^2_{e0})$ describes the variation between subjects in a cluster at a given time point; and $e_{1ij} \sim N(0, \sigma^2_{e1})$ (interaction of subject with time) describes the variation over time of a given subject. Note that an implicit sum is understood when indices are repeated, for example, $\gamma_{m00} t_m = \gamma_{100} t_1 + \ldots \gamma_{T00} t_T$ which will be γ_m whenever it concerns the mth measurement.

Note that we assume a constant treatment effect over time (the coefficient of x_{jm} is constant over time), or otherwise stated, we are interested in an average effect over time. Due to the random effects, a constant correlation among repeated measures on a given unit is introduced (compound symmetry).

As in the cluster randomized crossover design, the treatment effect can be estimated from between-cluster comparisons and within-cluster comparisons. For example, in the simplest stepped wedge with three measurements, the treatment effect estimator can be seen as a weighted average of three comparisons: the difference in change from baseline to the first post-baseline measurement $(t = 1)$, the difference in change from the first to the second post-baseline measurement $(t = 1 \rightarrow 2)$, and the difference at the first post-baseline measurement (De Hoop, Moerbeek, Borsje, & Teerenstra, 2014). The first two are within-cluster comparisons and the last is a between-cluster comparison.

Content matter experts may more easily think of the outcome in terms of correlations: how similar are the outcomes of subjects within the same cluster (ICC)? How similar are outcome scores of the same subject (in a given cluster) at two different time points (subject autocorrelation)? How similar is a cluster (average) when it is measured twice over time (cluster autocorrelation)? Therefore, the sample size formula is more practical if it is expressed in the ICC

$$\rho = \frac{(\sigma_{u0}^2 + \sigma_{u1}^2)}{(\sigma_{u0}^2 + \sigma_{u1}^2) + (\sigma_{e0}^2 + \sigma_{e1}^2)},$$

the cluster autocorrelation

$$\rho_c = \frac{\sigma_{u0}^2}{\sigma_{u0}^2 + \sigma_{u1}^2},$$

and the subject autocorrelation

$$\rho_s = \frac{\sigma_{e0}^2}{\sigma_{e0}^2 + \sigma_{e1}^2}.$$

In terms of these correlations, both cross-sectional and cohort designs and mixtures can be captured. For simplicity, the discussion below will be on cross-sectional designs and more specifically, the case that $\rho_c = 1$ in combination with $\rho_s = 0$. This case arises, for example, whenever new subjects are included in the clusters ($\rho_s = 0$), such as in a cross-sectional design, while at the same time the clusters perform consistent over time in terms of the outcome. The variation over time of the average of a cluster can be explained totally by the variation due to sampling new subjects at each time point ($\rho_c = 1$). For this case, Hussey and Hughes (2007) provided the treatment effect estimator and power formula in terms of variance components. In terms of correlations, the *variance* of the treatment effect estimator is

$$var(\gamma_{001}) = \frac{4\sigma^2}{n_1 n_2} \cdot \frac{1 + (Tn_1 + n_1 - 1)\rho}{1 + (\frac{1}{2}Tn_1 + n_1 - 1)\rho} \cdot \frac{3(1-\rho)}{2(T - \frac{1}{T})}.$$

Here n_1 is the number of subjects in a cluster at a given measurement. As these subjects are replaced at each measurement, each cluster will have contained $(T+1)n_1$ subjects during the course of the trial. The factor $\frac{4\sigma^2}{n_1 n_2}$ is the variance of a posttest design for a simple randomized trial (with $n_2 n_1$ understood as the total number of subjects at a given measurement). This means that the sample size of a stepped wedge cluster randomized trial can be obtained by a two-step procedure. First calculate, for the desired level of power $(1 - \beta)$ and two-sided significance level (α), the sample size to detect a standardized treatment effect $\delta = \gamma_{001}/\sigma$ for an *individually randomized parallel group* trial so that $N_u = 4(z_{1-\alpha/2} + z_{1-\beta})^2)/\delta^2$. Then multiply this unadjusted sample size by the design effect

$$RE_{SW\|\rho_c=1,\rho_s=0} = \frac{1 + (Tn_1 + n_1 - 1)\rho}{1 + (\frac{1}{2}Tn_1 + n_1 - 1)\rho} \cdot \frac{3(1-\rho)}{2(T - \frac{1}{T})}; \qquad (5.28)$$

see Woertman et al. (2013).

Thus, the sample size of the cluster randomized stepped wedge trial *per measurement* is $N = RE_{SW}N_u$ and at the end of the trial $RE_{SW}N_u(T+1)$ subjects will have been in the trial. The required *total* number of clusters is $n_2 = RE_{SW}N_u/n_1$. Multiplying this by the small sample correction factor $(n_2+1)/(n_2-1)$ (approximately) accounts approximately for the loss of power due to estimation of variance components (see p. 108 of Steel and Torrie (1980)).

The above procedure is for complete stepped wedge designs (i.e., all clusters are measured at each time point) with one measurement per step and two levels. Extensions to multiple measurements in a step are provided in Woertman et al. (2013). Incomplete designs and three-level designs are discussed by Hemming, Lilford, and Girling (2015).

The stepped wedge design requires a lower sample size in terms of the number of clusters than the post-test only parallel group design (see Figure 5.6). This is partly due to taking repeated measurements and partly because between-cluster comparisons and within-cluster comparisons were made. The number of subjects needed may, however, be substantial in the cross-sectional design, because each step requires new subjects. Therefore, the stepped wedge is most useful if the number of available clusters is limited but not (so much) the number of subjects that can be recruited into the clusters during the course of the trial.

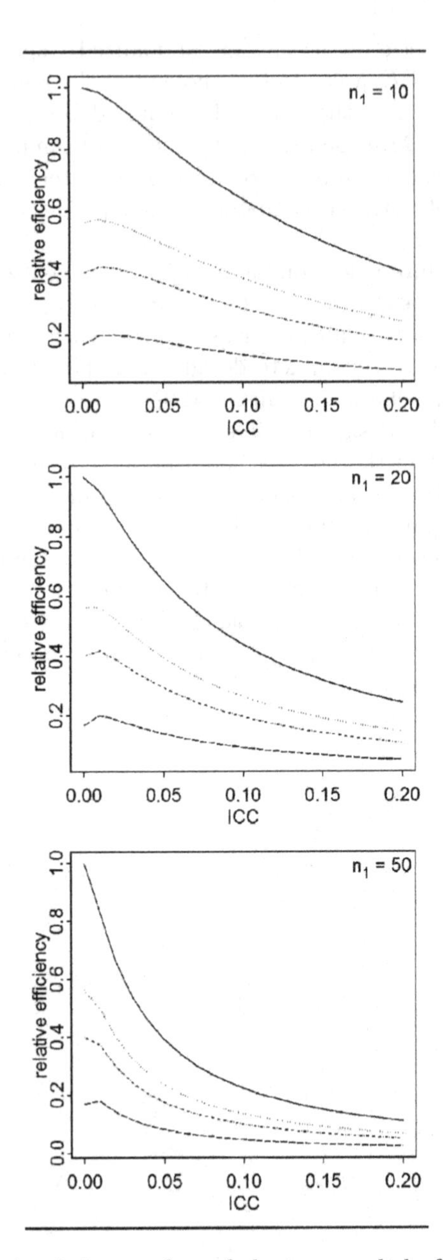

Figure 5.6: The ratio of the number of clusters needed of the stepped wedge versus the posttest design. From top to bottom, the solid lines represent T=2, dotted lines represent T=3, dot-dashed lines represent T=4, and dashed lines represent T=9.

6

Multisite trials

6.1 Introduction

A trial that is conducted in more than one center is generally referred to as a multisite, multicenter, or multiclinic trial. With such a trial, randomization to treatment conditions is done at the subject level within each center and more than one treatment is available within each center. Multisite trials are widely used in the medical sciences to compare different types of medication or surgery.

One of the major reasons to use a multisite trial is that any single cluster often cannot provide a sufficient number of subjects within reasonable time. For instance, a 3-month study might take 13 months to complete in a single center if the duration between accrual of the first and last patient is 10 months while it might only take 6 months to complete if multiple centers are involved and accrual is achieved in just 3 months. As is obvious, shorter study duration is related to lower costs and a reduction of lost sales of the sponsor of the trial. In addition, a reduction in travel costs is achieved when a multisite trial is used since it is generally cheaper to sample subjects who are geographically clustered than subjects who are widely dispersed.

Once the trial is concluded, data from all participating centers should be used in a statistical analysis to estimate the treatment effect and test for its significance. Had all centers conducted the trial on an individual basis and published the results in separate papers, the results of some of the centers might have been lost in the publication process. A meta-analysis that aims to combine the results of all individual centers might therefore have resulted in an overestimate of the treatment effect since the results from centers with non-significant treatment effects are more likely not to be included in a meta-analysis. This process is generally known as publication bias.

Despite its advantages, a multisite trial suffers from some drawbacks, the most important one being the risk of control group contamination. Such contamination occurs when information on the contents of the experimental condition leaks to the control condition and results in an underestimate of the treatment effect. It is unlikely to occur in double-blind pharmaceutical trials but very likely in trials that rely on interpersonal communication. This may be considered the major reason why multisite trials are less common in the social and behavioral sciences than in the medical sciences. In addition, a standardization of the delivery of the treatments over the participating clusters is required. This may not be that difficult to achieve in pharmaceutical studies but may be much more difficult in trials where treatments are delivered by professionals such as surgeons or therapists.

The effect of treatment may vary across the clusters that are involved in a multisite trial. In other words, there may be a treatment by cluster interaction. This is a property that is often considered both an advantage and disadvantage. On the one side, it would be helpful if one would be able to present one treatment effect value that holds for all clusters in the trial. On the other side, varying treatment effects allow the researcher to identify the clusters where the effect of treatment is largest and to investigate which cluster and patient characteristics explain the treatment effect difference. For instance, the highest treatment effects may be found in clinics with the best implementation of treatment, the best communication protocols between clinicians and patients and the most experienced professionals. These findings can then be used to improve treatment delivery in the other clinics.

Multisite trials can be shown to be more efficient than cluster randomized trials. That is, for any number of clusters of any cluster size, the standard error of the treatment effect estimate of a multisite is smaller than that of a cluster randomized trial. A higher power level can be achieved with a multisite trial, given that control group contamination is absent. Despite this advantage, less attention in the literature has been paid to multisite trials than to cluster randomized trials. A special issue in *Statistical Methods in Medical Research* appeared in 2005 (Jones, 2005). A series of three papers appeared in *Statistics in Medicine* in 1998 (Gould, 1998; Jones, Teather, Wang, & Lewis, 1998; Senn, 1998). In addition, attention to multisite trials has been paid in a few standard textbooks on clinical trials, for instance Friedman, Furberg, and DeMets (1999).

The introduction of Chapter 4 mentioned that clustering is not always correctly accounted for in the design and analysis of cluster randomized trials. This also appears to be the case for multisite trials (Biau, Porcher, & Boutron, 2008; Biau, Halm, et al., 2008). The aim of the present chapter is to present the correct statistical model for multisite trials and to provide sample size calculations. Sample sizes are given for trials where either the cluster size or the number of clusters is fixed, and for trials where a cost function is used to calculate the optimal allocation of units. Sample size calculations are given for continuous and dichotomous outcomes. An example in the last section shows the use of the results in a practical setting.

6.2 Multilevel model

The following hypothetical example is used to illustrate the use of the multilevel model to analyze data from a multisite trial. An anaesthetist wants to compare the effects of an experimental medication for pain relief to a standard type. Multiple clinics are enrolled in the trial and randomization to treatment conditions is done at the patient level within each clinic, so that both treatments are available within each clinic. The number of clinics is denoted n_2 and the number of patients within each clinic is denoted n_1. The cluster size is assumed not to vary across clinics. Randomization is done such that $\frac{1}{2}n_1$ patients per clinic receive the experimental treatment and the others receive the control; hence a balanced design is used.

Artificial data of such a trial with 20 clinics of size 10 are displayed in Table 6.1, which clearly shows that treatment is crossed with clinic. Hence, a multisite trial has a crossed design. Pain scores are recorded on a scale that ranges from 0 to 10; the lower the score, the lower the patient's perceived pain level. On average lower pain scores are observed in the experimental condition. The effect of treatment varies over clinics. For instance, the difference in treatment means is equal to $\mu_E - \mu_C = 1.2$ in clinic 1 and equal to $\mu_E - \mu_C = -0.8$ in clinic 2. So, the experimental condition performs worse than the control in clinic 1 and better than the control in clinic 2.

The multilevel regression model relates outcome score y_{ij} of patient i in clinic j to treatment condition x_{ij}. The model at the patient level is given by

$$y_{ij} = \beta_{0j} + \beta_{1j}x_{ij} + e_{ij}. \tag{6.1}$$

Treatment condition is coded -0.5 for the control condition and 0.5 for the experimental condition; thus β_{0j} and β_{1j} represent the mean score and (unstandardized) treatment effect in clinic j. These random effects are further modeled as an average component γ and clinic-dependent deviation u:

$$\begin{aligned} \beta_{0j} &= \gamma_{00} + u_{0j} \\ \beta_{1j} &= \gamma_{10} + u_{1j}. \end{aligned} \tag{6.2}$$

The random effect u_{0j} represents the difference between clinic j's mean score and the grand mean γ_{00}; similarly the random effect u_{1j} represents the difference between clinic j's treatment effect and the average treatment effect γ_{10}. These random effects are assumed to follow a normal distribution with zero mean and variances σ^2_{u0} and σ^2_{u1}; their covariance is denoted σ_{u01}. They are assumed to be independent of the random effect at the patient level e_{ij}. The latter random effect has a normal distribution with zero mean and variance σ^2_e.

The single-equation model is obtained from substitution of Model (6.2) in Model (6.1):

$$y_{ij} = \gamma_{00} + \gamma_{10}x_{ij} + u_{0j} + u_{1j}x_{ij} + e_{ij}. \tag{6.3}$$

Table 6.1: Example data for multisite trial

clinic number	condition: experimental					control				
1	3	4	6	8	4	2	3	5	4	5
2	1	4	4	5	4	4	4	4	4	6
3	1	2	3	5	5	2	4	5	6	7
4	8	2	2	3	6	2	4	8	7	9
5	3	3	2	5	5	4	2	2	3	4
6	4	6	8	9	9	3	3	4	5	7
7	4	4	4	4	5	4	5	6	7	8
8	3	3	4	3	3	6	6	6	6	7
9	6	3	5	5	7	3	4	4	6	7
10	3	4	4	4	8	8	9	9	8	9
11	3	2	2	2	4	3	8	3	8	8
12	1	3	3	4	9	2	6	6	6	6
13	6	7	6	7	8	6	6	9	9	9
14	1	3	5	5	9	5	4	6	7	8
15	2	1	2	2	3	2	4	3	5	6
16	5	8	9	9	9	5	7	8	9	9
17	2	3	3	3	6	4	4	6	6	8
18	3	3	5	5	6	3	3	4	5	9
19	4	3	5	4	4	3	4	5	5	8
20	4	5	6	6	6	5	6	7	8	9

Table 6.2 shows the parameter estimates of Model (6.3) that are obtained for the data in Table 6.1 and RIGLS estimation. The treatment effect estimate is equal to -1.090 and the null hypothesis of no treatment effect, $H_0 : \gamma_1 = 0$, was tested by the Wald test. The test statistic $t = \hat{\gamma}_1/s.e.(\hat{\gamma}_1)$ follows a t distribution with $n_2 - 1$ degrees of freedom under the null hypothesis. For this particular data set, the p-value is 0.007, so the null hypothesis is rejected at the $\alpha = 0.05$ level and it is concluded that on average an effect of treatment exists.

An analysis that ignores the nested data structure would have produced the same point estimate for the treatment effect but a smaller standard error (s.e. $= 0.298$) and hence a smaller p-value of 0.000. In this case this would have resulted in the same conclusion with respect to the significance of the treatment effect at the $\alpha = 0.05$ level, but for other studies the conclusion may be different.

For the calculation of the intraclass correlation coefficient, a restriction is made to the multilevel model where $\sigma_{u01} = 0$, since that assumption also applies to the mixed effects ANOVA model (see below). Given coding $x_{ij} = -0.5$ for the control condition and $x_{ij} = 0.5$ for the intervention condition, the total variance of the outcome is

$$var(y_{ij}) = \sigma^2_{u0} + \frac{1}{4}\sigma^2_{u1} + \sigma^2_e. \tag{6.4}$$

Table 6.2: Results based on the multilevel model

parameter	estimate	standard error	t	p
fixed effects:				
intercept, γ_{00}	5.005	0.274		
slope, γ_{10}	-1.090	0.361	-3.023	0.007
random effects:				
intercept, σ^2_{u0}	1.206	0.488		
slope, σ^2_{u1}	1.418	0.854		
intercept slope covariance, σ_{u01}	0.118	0.454		
error term, σ^2_e	2.955	0.330		

The first two terms at the right side are clinic level variances and the third term is a variance at the patient level. The intraclass correlation coefficient is defined as the proportion variance at the clinic level:

$$\rho = \frac{\sigma^2_{u0} + \frac{1}{4}\sigma^2_{u1}}{\sigma^2_{u0} + \frac{1}{4}\sigma^2_{u1} + \sigma^2_e}. \tag{6.5}$$

As in previous chapters, the intraclass correlation coefficient varies between 0 and 1, and larger values indicate a larger degree of correlation among outcomes of patients within the same clinic. The intraclass correlation coefficient may be split in two parts that reflect the proportion variance due to the random clinic effect (ρ_0) and random clinic by treatment interaction (ρ_1). So, $\rho = \rho_0 + \rho_1$, where $\rho_0 = \sigma^2_{u0}/(\sigma^2_{u0} + \frac{1}{4}\sigma^2_{u1} + \sigma^2_e)$ and $\rho_1 = \frac{1}{4}\sigma^2_{u1}/(\sigma^2_{u0} + \frac{1}{4}\sigma^2_{u1} + \sigma^2_e)$. For the model in Table 6.2 it is calculated that $\rho = 0.346$ and $\rho_0 = 0.267$, $\rho_1 = 0.078$.

A value ρ_1 close to zero suggests that the treatment effect does not vary across clinics. A deviance-based test should be applied to test whether this is indeed the case. The deviance for the model in Table 6.2 is equal to 831.438. Fitting a model without a random slope results in a deviance of 837.982. It should be noted that this latter model does not only lack the slope variance, but also lacks the covariance between intercept and slope. Thus, the difference in number of parameters between both models is equal to 2. The difference in deviances is equal to 6.544, and the test statistic follows a chi-squared distribution with two degrees of freedom under the null hypothesis of zero slope variance and intercept by slope covariance. The (halved) p-value is 0.020 and it is concluded that the null hypothesis is rejected. In other words, the treatment effect varies significantly across clinics.

Under certain restrictions, the multilevel model gives the same results as the mixed effects ANOVA model (Raudenbush, 1993). The ANOVA model for outcome y_{ijk} of patient k in clinic j in treatment condition i is written as

$$y_{ijk} = \mu + \alpha_i + \beta_j + (\alpha\beta)_{ij} + r_{ijk}. \tag{6.6}$$

The grand mean is denoted μ and the fixed effect of treatment i is denoted α_i. To be able to estimate the latter effect, the constraint $\sum_i \alpha_i = 0$ is used;

otherwise the two treatment effects α_i and the grand mean μ cannot be estimated simultaneously. The random clinic effect is denoted β_j and $(\alpha\beta)_{ij}$ is the random clinic by treatment interaction. These random effects are independently and normally distributed with zero mean and variances σ^2_α and $\sigma^2_{\alpha\beta}$. They are assumed to be independent from the patient level random effect $r_{ijk} \sim N(0, \sigma^2_r)$. It should be noted that the mixed effects ANOVA model assumes the clinic and clinic by treatment interaction to be independent. This is not the case for the multilevel model where the covariance σ_{u01} between the random intercept and slope is not forced to be equal to zero. However, when the design is balanced and the constraint $\sum_i \alpha_i = 0$ is used, this covariance does not have an effect on estimation and testing of fixed effects and variance components.

In principle there exist two possibilities for the mixed effects ANOVA model: one without restrictions on the interaction terms $(\alpha\beta)_{ij}$ and one with the restriction $\sum_i(\alpha\beta)_{ij} = 0$ for all j. This chapter assumes the latter since it also applies to the multilevel model. For the multilevel model, $\sum_i x_{ij}u_{1j} = 0$ for all j as there are only two treatments which are coded -0.5 and 0.5 and there are $\frac{1}{2}n_1$ patients per treatment per clinic. For more detailed comments on restrictions in ANOVA models the reader is referred to Pages 123 through 126 of Searle et al. (1992).

The degrees of freedom, sums of squares and expected mean squares are given in Table 6.3. The variance components are estimated from the mean squares as follows:

$$\hat{\sigma}^2_\beta = \frac{MS_{clinic} - MS_{error}}{n_1}, \; \hat{\sigma}^2_{\alpha\beta} = \frac{MS_{interaction} - MS_{error}}{n_1}$$
$$\text{and } \hat{\sigma}^2_r = MS_{error}. \tag{6.7}$$

These estimates may be negative and common practice is to replace negative estimates by the value zero. The test statistic for the null hypothesis of no overall treatment effect is given by $F = MS_{treatment}/MS_{interaction}$. Under the null hypothesis, this test statistic follows an F distribution with 1 and $n_2 - 1$ degrees of freedom. In case one wishes to test if the treatment effect varies over clinics one should use the test statistic $F = MS_{interaction}/MS_{error}$. This test statistic has an F distribution with $n_2 - 1$ and $n_1n_2 - 2n_2$ degrees of freedom under the null hypothesis of no clinic by treatment interaction.

If $i = 1$ for the control treatment and $i = 2$ for the experimental treatment then the parameters in the multilevel model (6.3) and ANOVA model (6.6) are related as follows:

$$\mu = \gamma_{00}, \; \alpha_2 - \alpha_1 = \gamma_{10}, \; \beta_j = u_{0j}, \; (\alpha\beta)_{2j} - (\alpha\beta)_{1j} = u_{1j}, \; r_{ijk} = e_{ij}. \tag{6.8}$$

Table 6.4 shows the parameter estimates for the mixed effects ANOVA model. The grand mean and average unstandardized treatment effect are estimated as $\hat{\mu} = \overline{y}_{...} = 5.005$ and $\hat{\alpha}_2 - \hat{\alpha}_1 = \overline{y}_{2..} - \overline{y}_{1..} = 4.460 - 5.550 = -1.090$. These values correspond to the grand mean and treatment effect of the multilevel model. The F test statistic is equal to the square of the t test statistic

Table 6.3: Analysis of a crossed mixed effects ANOVA model

Source	*df*	*SS*	$E(MS)$
treatment	1	$\frac{1}{2}n_1n_2\sum_i(\overline{y}_{i..}-\overline{y}_{...})^2$	$\sigma_r^2+n_1\sigma_{\alpha\beta}^2+\frac{1}{2}n_1n_2\Sigma_i\alpha_i^2$
clinic	n_2-1	$n_1\sum_j(\overline{y}_{.j.}-\overline{y}_{...})^2$	$\sigma_r^2+n_1\sigma_\beta^2$
interaction	n_2-1	$n_1\sum_i\sum_j(\overline{y}_{ij.}-\overline{y}_{i..}-\overline{y}_{.j.}-\overline{y}_{...})^2$	$\sigma_r^2+n_1\sigma_{\alpha\beta}^2$
error	$n_1n_2-2n_2$	$\sum_i\sum_j\sum_k(y_{ijk}-\overline{y}_{ij.})^2$	σ_r^2
corrected total	n_1n_2-1	$\sum_i\sum_j\sum_k(y_{ijk}-\overline{y}_{...})^2$	

Table 6.4: Results based on the mixed effects ANOVA model

Source	*df*	*SS*	*MS*	*F*	*p*
treatment	1	59.405	59.504	9.140	0.007
clinic	19	285.295	15.016	2.310	0.038
interaction	19	123.495	6.500	2.200	0.004
error	160	472.800	2.955		
corrected total	199	940.995			

for the multilevel model. As for the multilevel model the p-value is equal to 0.007 and the null hypothesis of no treatment effect is rejected at the $\alpha = 0.05$ level.

On the basis of Equation (6.7), the variance components are estimated as $\hat{\sigma}^2_\beta = 1.206$, $\hat{\sigma}^2_{\alpha\beta} = 0.3545$ and $\hat{\sigma}^2_r = 2.955$. The variance component for the treatment by clinic interaction is four times smaller than the variance of the random treatment effect in the multilevel model. The estimates for the other two variances correspond to their counterparts in the multilevel model. From Table 6.4 it is observed that both the clinic and treatment by clinic interaction are significant. In other words, the mean score and treatment effect vary across clinics.

Constant treatment effect

The multilevel model and mixed effects ANOVA model discussed thus far assume the effect of treatment varies over clinics. This is expressed by the interaction between treatment and clinic in both models. If this interaction turns out to be non-significant, it may be removed from the model. The multilevel model then simplifies to

$$y_{ij} = \gamma_{00} + \gamma_{10}x_{ij} + u_j + e_{ij}. \tag{6.9}$$

The variance of the outcome y_{ij} is equal to $var(y_{ij}) = \sigma^2_u + \sigma^2_e$. The corresponding intraclass correlation coefficient is $\rho = \sigma^2_u/(\sigma^2_u + \sigma^2_e)$. The corresponding mixed effects ANOVA model becomes

$$y_{ijk} = \mu + \alpha_i + \beta_j + r_{ijk}. \tag{6.10}$$

The sums of squares and expected mean squares are shown in Table 6.5. It is observed that the sums of squares and degrees of freedom for the interaction effect are added to those for the error term. The $E(MS_{treatment})$ does not depend on $\sigma^2_{\alpha\beta}$, since this variance is equal to zero in a model with a zero treatment by clinic interaction effect. The test statistic for the test on treatment effect is equal to $F = MS_{treatment}/MS_{error}$ and follows an F distribution with 1 and $n_1n_2 - n_2 - 1$ degrees of freedom under the null hypothesis of no treatment effect.

As will be shown in the next section, the presence or absence of a treatment by clinic interaction determines the relation between sample size and statistical power.

Table 6.5: Analysis of a crossed mixed effects ANOVA model without treatment by clinic interaction

Source	*df*	*SS*	*E(MS)*
treatment	1	$\frac{1}{2}n_1n_2\sum_i(\overline{y}_{i..}-\overline{y}_{...})^2$	$\sigma_r^2+\frac{1}{2}n_1n_2\Sigma_i\alpha_i^2$
clinic	n_2-1	$n_1\sum_j(\overline{y}_{.j.}-\overline{y}_{...})^2$	$\sigma_r^2+n_1\sigma_\beta^2$
error	$n_1n_2-n_2-1$	$\sum_i\sum_j\sum_k(y_{ijk}-\overline{y}_{i..}-\overline{y}_{.j.}-\overline{y}_{...})^2$	σ_r^2
corrected total	n_1n_2-1	$\sum_i\sum_j\sum_k(y_{ijk}-\overline{y}_{...})^2$	

6.3 Sample size calculations for continuous outcomes

This section follows the same structure as Section 4.3 for cluster randomized trials. The non-centrality parameter of the test statistic for the treatment effect is presented first and we show how it depends on the variance components, cluster size, number of clusters and effect size. The design effect determines how often a multisite trial should be replicated to perform as well as a simple randomized trial without clustering of subjects within clusters. Attention is paid to sample sizes when either the number of clusters or the cluster size is fixed. The situation where neither sample size is fixed and a budgetary constraint is used to calculate the optimal allocation of units is also discussed.

The sample size formulae in this section are based on Parzen, Lipsitz, and Dear (1998), Vierron and Giraudeau (2007), Vierron and Giraudeau (2009), Moerbeek et al. (2000) and Raudenbush and Liu (2000). The first four papers focus on trials where the treatment effect is constant across clusters; the latter paper pays attention to the case where each cluster has its own treatment effect.

6.3.1 Factors that influence power

The equations presented in this section assume a balanced design with clusters having equal size n_1 and equal allocation to treatment conditions within each cluster. In other words, $\frac{1}{2}n_1$ subjects per cluster are assigned to the control condition and the other $\frac{1}{2}n_1$ subjects are assigned to the experimental condition.

To test the null hypothesis of no treatment effect, $H_0 : \gamma_{10} = 0$, the test statistic $t = \gamma_{10}/s.e.(\gamma_{10})$ is used. Under the null hypothesis this test statistic follows a central t distribution with n_2-1 degrees of freedom. Under the alternative hypothesis it follows a non-central t distribution with non-centrality parameter

$$\lambda = \frac{\gamma_{10}}{\sqrt{\frac{4(\sigma_e^2+\frac{1}{4}n_1\sigma_{u1}^2)}{n_1n_2}}}. \tag{6.11}$$

The numerator of the non-centrality parameter is the unstandardized treatment effect γ_{10}, the denominator is the standard error of the treatment effect estimator:

$$s.e.(\hat{\gamma}_{10}) = \sqrt{\frac{4(\sigma_e^2 + \frac{1}{4}n_1\sigma_{u1}^2)}{n_1 n_2}}. \tag{6.12}$$

One may note the similarity between this standard error and the standard error for a cluster randomized trial as given by Equation (4.8). The only difference is that σ_u^2 in Equation (4.8) is replaced by $\frac{1}{4}\sigma_{u1}^2$ in Equation (6.12). As $\frac{1}{4}\sigma_{u1}^2 < \sigma_u^2$, a multisite trial is more efficient than a cluster randomized trial.

Given total variance $\sigma^2 = \sigma_e^2 + \sigma_{u0}^2 + \frac{1}{4}\sigma_{u1}^2$ the non-centrality parameter may be expressed in terms of the standardized treatment effect $\delta = \gamma_{10}/\sigma$size and the two components ρ_0 and ρ_1 of the intraclass correlation coefficient ρ:

$$\lambda = \frac{\delta}{\sqrt{\frac{4(1-\rho_0+(n_1-1)\rho_1)}{n_1 n_2}}}. \tag{6.13}$$

The power for the test on treatment effect is calculated from

$$\begin{array}{r} P(t_{n_2-1,\lambda} > t_{n_2-1,1-\alpha}) \text{ when } H_a: \gamma_{10} > 0 \\ P(t_{n_2-1,\lambda} < t_{n_2-1,\alpha}) \text{ when } H_a: \gamma_{10} < 0 \\ P(t_{n_2-1,\lambda} > t_{n_2-1,1-\alpha/2}) + P(t_{n_2-1,\lambda} < t_{n_2-1,\alpha/2}) \text{ when } H_a: \gamma_{10} \neq 0 \end{array}. \tag{6.14}$$

As for any statistical test, the power level increases with increasing type I error rate α. Thus the effects of α on power are not studied further in the remainder of this chapter. All figures in this section show power curves to detect a small treatment effect in a test with a two-sided alternative hypothesis at the common value $\alpha = 0.05$.

Equation (6.13) shows that the non-centrality parameter, and hence the statistical power, depends on the number of clusters n_2, cluster size n_1, intraclass correlation coefficient components ρ_0 and ρ_1 and effect size δ. The effects of these factors are now studied in further detail.

Number of clusters. The number of clusters only appears in the denominator of the standard error (6.12). This implies that the standard error decreases to zero, and hence the power increases to 100%, when the number of clusters increases to infinity. Figure 6.1 shows the relation between number of clusters and power to detect a small treatment effect. The two lines within each graph represent different cluster sizes $n_1 = 10$ and $n_1 = 20$. The left graph corresponds to a model where the effect of treatment is constant across clusters (i.e., $\rho_1 = 0$). The right graph assumes a random treatment effect across clusters and the corresponding variance component is assumed to be 5% of the total variance (i.e., $\rho_1 = 0.05$). In both graphs the variance of the random intercept is assumed to be 5% of the total variance (i.e., $\rho_0 = 0.05$).

It is observed that power increases with increasing cluster size and increasing numbers of clusters. Higher power levels are achieved if the treatment

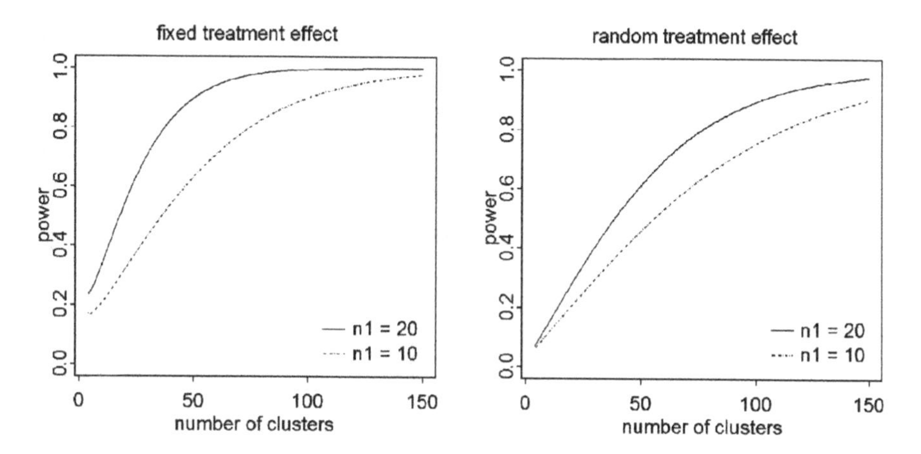

Figure 6.1: Power as a function of the number of clusters.

effect does not vary across clusters, which is explained by the fact that the term $n_1\sigma_{u1}^2$ in the numerator of the standard error (6.12) is equal to zero if the treatment effect is constant.

Cluster size. Figure 6.2 shows that the power to detect a small treatment effect increases with increasing cluster size. Increasing cluster size to a large value will not always be sufficient to achieve a desired power level in case the treatment effect varies across clusters. This is depicted in the right graph of Figure 6.2 where both the intercept and slope variance are 5% of the total variance ($\rho_0 = \rho_1 = 0.05$). Increasing cluster size will result in a power of less than 60% when only 30 clusters are involved in the trial. One strategy to increase power is to use a larger number of clusters. For a trial with 60 clusters, a power level of about 90% can be achieved.

The finding that increasing cluster size alone might not be sufficient to achieve a desired power is explained by the fact that cluster size appears in both the numerator and denominator of the standard error of the treatment effect estimator; see Equation (6.12). A standard error equal to $\sqrt{\sigma_{u1}^2/n_2}$ is achieved when the cluster size approaches infinity, and this value is only equal to zero when $\sigma_{u1}^2 = 0$. In other words, in a trial with a small number of clusters, a power level of 100% can only be achieved when the treatment effect does not vary across clusters. This is depicted in the left graph of Figure 6.2 where power increases to 100% even when the number of clusters is as small as 30. In this graph the slope variance is equal to zero and the intercept variance is 5% of the total variance ($\rho_0 = 0.05$; $\rho_1 = 0$).

Intraclass correlation coefficient. The non-centrality parameter (6.13) depends on the intraclass correlation coefficients ρ_0 and ρ_1: higher values ρ_0 result in higher power levels but higher values ρ_1 result in lower power levels. The effect

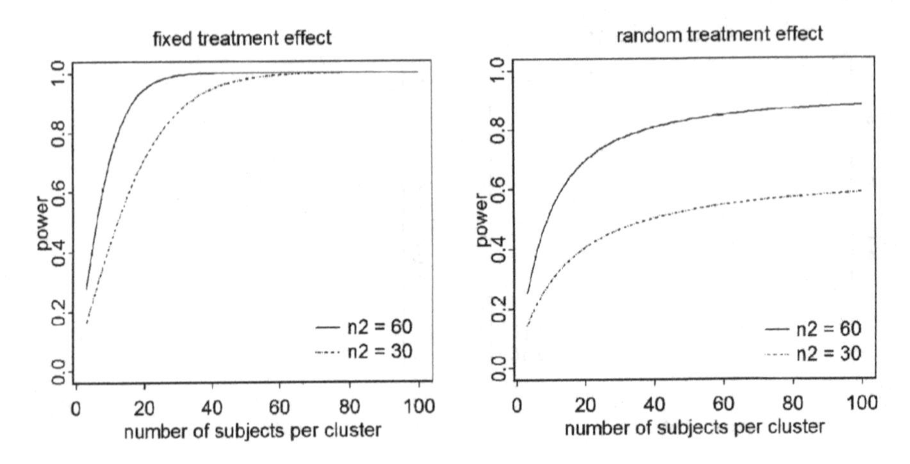

Figure 6.2: Power as a function of cluster size.

of ρ_1 is more pronounced than that of ρ_0, since it is multiplied by $n_1 - 1$. Figure 6.3 shows the relation between ρ_1 and power for a trial with 30 clusters of 30 subjects each. Power levels are shown for a small, medium and large effect size δ. For any effect size, the power decreases if ρ_1 increases to larger values. For a small effect size the decline in power is largest for small values of ρ_1 and for a large effect size the steepest decline can be found for larger values of ρ_1. In theory, the value of ρ_1 ranges from 0 to 1 but in practice much smaller values of, say, less than 0.20 are observed. Within the range $[0, 0.2]$ the effect of increasing ρ_1 on power is largest for a small treatment effect and negligible for a large treatment effect. Figure 6.3 is based on the assumption that $\rho_0 = 0.05$. Changing this value hardly influences the power levels and graphs for other values of ρ_0 are therefore not given here.

Effect size. As for any statistical test the power increases with increasing effect size. The right side of Figure 6.3 shows the power as a function of the standardized effect size δ and for different values of ρ_1, given $\rho_0 = 0.05$ and 30 clusters of size 30. It is observed that power increases with increasing effect size, and higher power levels are achieved for smaller ρ_1. For higher values ρ_1, larger standardized effect sizes are needed to gain the same level of power (e.g., 0.8) than for smaller values of ρ_1. This implies that a higher sample size is required for higher treatment effect variability. Figure 6.3 was made under the assumption that $\rho_0 = 0.05$. Similar graphs were found for other values ρ_0 and are not given here.

6.3.2 Design effect

The design effect determines how often a multisite trial should be replicated to perform as well as a simple randomized trial that does not involve nesting

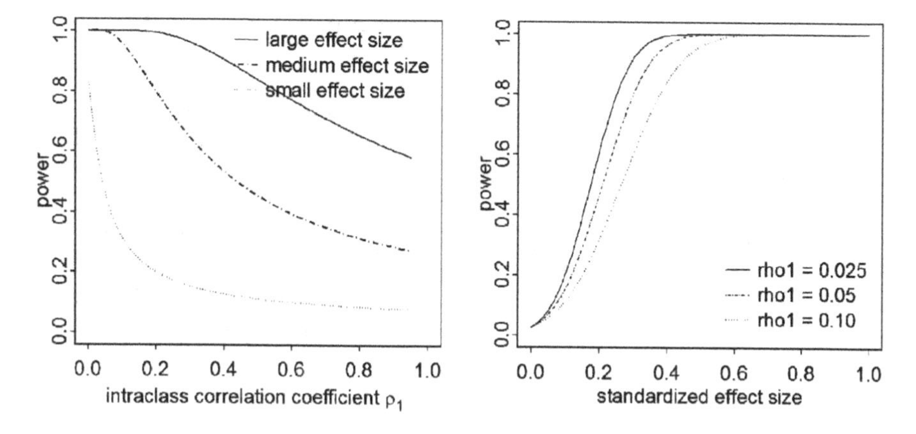

Figure 6.3: Power as a function of the intraclass correlation coefficient and effect size.

of subjects within clusters. For a multisite trial the design effect is given by

$$\text{design effect} = 1 - \rho_0 + (n_1 - 1)\rho_1. \tag{6.15}$$

The value of the design effect depends on the cluster size n_1 and the two components ρ_0 and ρ_1 of the intraclass correlation coefficient. The value is larger than 1 when $\rho_1 > \rho_0/(n_1 - 1)$, meaning that a multisite trial performs worse than a simple randomized trial in terms of statistical power if the variance of the random slope is sufficiently large.

When the treatment effect is constant across clusters, the design effect reduces to $1 - \rho_0$ and, irrespective of the cluster size, the multisite trial performs better than the simple randomized trial. A cluster serves as its own control since both treatment conditions are available within each cluster and the between-cluster variability of the outcome is not taken into account in the calculation of the standard error of the treatment effect estimator. It should be noted that a similar result was found for crossover trials (Section 5.4). For such trials, both treatment conditions are available within a subject and hence each subject serves as its own control. This implies that the standard error of the treatment effect estimator does not depend on the between-subject variance in a crossover trial.

The design effect of a multisite trial is different from the design effect of a cluster randomized trial. In principle, each design has its own design effect. It is therefore important to carefully select the appropriate design before the required sample sizes at the cluster and subject level are calculated.

6.3.3 Sample size formulae for fixed cluster size or fixed number of clusters

To calculate the required sample size, a decision should be made on the desired power level $1-\beta$, the type I error rate α and the unstandardized or standardized effect size γ_{10} or $\delta = \gamma_{10}/\sigma$. Once their values are selected, the total cluster size n_1 and number of clusters n_2 can be calculated.

In some cases the cluster size is fixed in advance. For instance, the number of patients that can be treated by a therapist or the number of pupils per class may be fixed to some value at the beginning of the trial. When all clusters have the same size n_1, the number of clusters to achieve adequate power is calculated as

$$\begin{aligned} n_2 &= \frac{4(\sigma_e^2 + \frac{1}{4}n_1\sigma_{u1}^2)}{n_1}\left(\frac{z_{1-\alpha}+z_{1-\beta}}{\gamma_{10}}\right)^2 \\ &= \frac{4(1-\rho_0+(n_1-1)\rho_1)}{n_1}\left(\frac{z_{1-\alpha}+z_{1-\beta}}{\delta}\right)^2. \end{aligned} \tag{6.16}$$

For a test with a two-sided alternative hypothesis the term α is replaced by $\alpha/2$. In practice the number of clusters as calculated from this formula is a non-integer value and has to be rounded upward to the nearest integer. The achieved power will then be slightly larger than the desired power level.

Equation (6.16) assumes equal cluster size n_1, while in many trials cluster sizes vary to some degree. Van Breukelen et al. (2007) studied the relative efficiency of multisite trials with equal and unequal cluster sizes. They concluded that the loss of efficiency of using unequal cluster sizes rarely exceeds 10% and this can be compensated by including 11% more clusters in the trial. This implies that the number of clusters as calculated from Equation (6.16) needs to be multiplied by 1.11 to achieve the desired power level.

It should be noted that the normal approximation to the t distribution was used to calculate the required number of clusters. An explicit formula for the required number of clusters cannot be derived on the basis of the t distribution since the degrees of freedom of the t distribution are a function on the number of clusters. It is often argued that the difference between the t distribution and its normal approximation becomes negligible when the degrees of freedom exceed 30, with $df = n_2 - 1$ for a multisite trial. For fewer clusters one could use Equation (6.14) to calculate the relation between the number of clusters and power for a large range of numbers of clusters and select the smallest number for which the desired power level is achieved.

In some trials the number of clusters is fixed in advance. For instance, the number of therapists or general practices that are willing to participate in the trial might be limited. In that case the number of subjects per cluster is calculated as

$$n_1 = \frac{4\sigma_e^2}{\left(\frac{\gamma_{10}}{z_{1-\alpha}+z_{1-\beta}}\right)^2 n_2 - \sigma_{u1}^2} = \frac{4(1-\rho_0-\rho_1)}{\left(\frac{\delta}{z_{1-\alpha}+z_{1-\beta}}\right)^2 n_2 - 4\rho_1}. \tag{6.17}$$

As usual, α is replaced by $\alpha/2$ when a statistical test with a two-sided alternative hypothesis is performed. The cluster size obtained from this equation should be rounded upward to the nearest even integer, since equal allocation to treatment conditions within each cluster is assumed.

Unequal allocation ratios. A balanced design with 50:50 allocation of subjects within clusters to treatment conditions may not always be the optimal allocation. Unequal allocation ratios may be more efficient, for instance, when the costs per subject in the one condition are much higher than those in the other condition. Liu (2003) studied the case where the costs at the subject level varied across treatment conditions. He found that the optimal allocation ratio r of subjects to treatment conditions within each cluster is equal to $r = n_{1_E}/n_{1_C} = \sqrt{c_C/c_E}$, where n_{1_E} and n_{1_C} are the number of subjects per cluster within the experimental and control conditions and c_E and c_C are the costs per subject in the experimental and control conditions.

The ratio of the costs at the subject level to achieve a desired power level for design with equal and unequal allocation is

$$\frac{Costs_{unequal}}{Costs_{equal}} = \frac{(1+r)(c_E + c_C/r)}{2(c_C + c_E)}. \tag{6.18}$$

This cost ratio was also found in Chapter 4 for cluster randomized trials. In that chapter it was also shown that substantial reductions in the costs of 10% or higher can be achieved when the cost ratio is 4 or larger.

6.3.4 Including budgetary constraints

When neither the cluster size nor the number of clusters is fixed beforehand, a budgetary constraint may be used to calculate the required sample sizes at the cluster and subject levels. Suppose the costs to include a cluster c_2 and the costs to sample an additional subject per cluster are c_1, respectively. The non-overhead costs of the trial are calculated by multiplying the number of clusters by the costs at the cluster level and adding the total costs at the subject level, which are calculated as the product of the total number of subjects and the costs at the subject level. These costs should not exceed the budget C that is available for including clusters and subjects in the trial:

$$c_1 n_1 n_2 + c_2 n_2 \leq C. \tag{6.19}$$

The average of the costs at the subject level should be used in the budgetary constraint if these costs vary across treatments. This is permitted since equal numbers of subjects per treatment per cluster are assumed in the calculations that follow.

A maximal power given the budgetary constraint (6.19) is desired. In other words, the standard error (6.12) should be minimized such that the costs do not exceed the budget. To calculate the optimal allocation of units, the budgetary constraint is used to express the number of clusters n_2 in terms of the budget C, costs c_1 and c_2 and cluster size n_1

$$n_2 = C/(c_1 n_1 + c_2). \tag{6.20}$$

This expression is substituted into the standard error (6.12) and the optimal cluster size n_1^* is derived. This value is then substituted into Equation (6.20) to calculate the optimal number of clusters n_2^*.

Following this procedure the optimal cluster size is

$$n_1^* = \sqrt{\frac{c_2 \sigma_e^2}{c_1 \frac{1}{4} \sigma_{u1}^2}} = \sqrt{\frac{c_2(1-\rho_0-\rho_1)}{c_1 \rho_1}}, \tag{6.21}$$

and the optimal number of clusters is equal to

$$n_2^* = \frac{C}{\sqrt{\frac{c_1 c_2 \sigma_e^2}{\frac{1}{4}\sigma_{u1}^2}} + c_2} = \frac{C}{\sqrt{\frac{c_1 c_2 (1-\rho_0-\rho_1)}{\rho_1}} + c_2}. \tag{6.22}$$

A lager cluster size, and hence a smaller number of clusters, is needed when the cost ratio c_2/c_1 increases. In other words, more subjects per cluster and fewer clusters are required when it becomes more expensive to include a cluster relative to including a subject. In addition to that, increasing the budget C has an effect on the number of clusters, but not on the cluster size. It should be mentioned that similar results were found for cluster randomized trials in Chapter 4. The only difference is that a cluster randomized trial uses σ_u^2 while a multisite trial uses $\frac{1}{4}\sigma_{u1}^2$ in the optimal sample size equations.

The standard error of the treatment effect estimator for the optimal sample sizes is equal to

$$s.e.(\hat{\gamma}_{10}) = \frac{2\sqrt{c_1 \sigma_e^2} + \sqrt{c_2 \sigma_{u1}^2}}{\sqrt{C}} = 2\sigma \frac{\sqrt{c_1(1-\rho_0-\rho_1)} + \sqrt{c_2 \rho_1}}{\sqrt{C}}. \tag{6.23}$$

As the sample sizes in Equations (6.21) and (6.22) are rounded to integer values, the standard error becomes somewhat larger. As is obvious, a smaller standard error is achieved when the budget C increases.

6.3.5 Constant treatment effect

Equations (6.21) through (6.23) apply to trials with a treatment effect that varies across clusters. The optimal design is different if one expects the treatment effect to be constant, which implies the data are generated from Equation (6.9). In that case the variance component corresponding to the treatment by cluster interaction σ_{u1}^2 is set to zero in the standard error of the treatment effect estimator and Equation (6.12) becomes

$$s.e.(\hat{\gamma}_{10}) = \sqrt{\frac{4\sigma_e^2}{n_1 n_2}}. \tag{6.24}$$

The optimal design is found by substituting the number of clusters n_2 from Equation (6.20) into the standard error (6.24) and solving for the cluster size n_1. The optimal design includes just one cluster and spends all remaining budget on including as many subjects as possible within that single cluster. The problem with such a design is that a single cluster cannot be regarded as a representative sample from the underlying population of clusters. A multilevel analysis is therefore not justified and even not possible since the cluster level variance σ_u^2 cannot be estimated with a single cluster. A rule of thumb by Snijders and Bosker (2012) states that at least 20 clusters are needed to perform a multilevel analysis. A design with 20 clusters allows us to sample

$$n_1 = \frac{C - 20c_2}{20c_1} \tag{6.25}$$

subjects per cluster given the budget constraint (6.19). The corresponding standard error is calculated as

$$s.e.(\hat{\gamma_{10}}) = 2\frac{\sqrt{c_1\sigma_e^2}}{\sqrt{C - 20c_2}} = 2\sigma\frac{\sqrt{c_1(1-\rho)}}{\sqrt{C - 20c_2}}, \tag{6.26}$$

where $\rho = \sigma_u^2/(\sigma_u^2 + \sigma_e^2)$. The actual standard error will generally be slightly larger due to rounding of sample sizes to integer values.

It should be mentioned that a design with as few as 20 clusters does not allow an accurate estimate of the cluster level variance component (Maas & Hox, 2005). This should no be a problem if the main objective of the trial is to estimate the treatment effect and its significance, since the standard error of the treatment effect estimator does not depend on the cluster level variance. If the estimation of cluster level random effect and its variance σ_u^2 is of importance, for instance to rank the clusters with respect to their performance, then more than 20 clusters are required.

Equation (6.24) shows that the standard error of the treatment effect estimator does not depend on the variance component at the cluster level. This is also true for the cluster randomized cluster crossover trial in case the correlation between outcomes of two subjects in the same cluster does not depend on whether these subjects are from the same or different time periods, see Equation (5.14) in Section 5.4. Thus, these two designs are as efficient with respect to estimating the treatment effect but they show differences with respect to the number of periods.

With a multisite trial, one single period is used and all treatment conditions are administered in the same period. With a cluster randomized cluster crossover trial, multiple periods are distinguished and one treatment per period per cluster is administered. The advantage of the multisite trial is that, since it is based on one period, it has shorter trial duration and hence study results become available sooner. The advantage of the cluster randomized cluster crossover trial is that, since one treatment is available per cluster per period, fewer mistakes may be made by the person who administers the treatments to the subjects. An additional advantage might be that since the trial has longer duration, more subjects may be recruited.

Consider as an example a trial conducted in Dutch hospitals to compare the effects of two treatment conditions and a control on the prevention of ventilator-associated pneumonia in patients within intensive care units (ICU) (De Smet et al., 2009). The first treatment focused on decontamination of the mouth only, whereas the second also focused on decontamination of the digestive tract. Thirteen hospitals were randomly assigned to different orders of the three treatments and different sets of ICU patients were used in different periods. The trial followed a cluster randomized cluster crossover design since crossover was done at the cluster level but not at the patient level within clusters.

The advantage of this trial over the multisite trial is that fewer mistakes in administering the treatments to the patients might have been made since only one treatment per ICU per period was used. In addition, recruitment of eligible patients might have been easier than in a multisite trial because of longer study duration. In total about 2000 patients per treatment could be recruited over the three periods.

Had a multisite trial with only one period been used, the researchers might have been able to recruit about a third of this number of patients. A longer study period may be a disadvantage over the crossover trial. Each period lasted 6 months and four washout periods of 1 month were used. Thus, each ICU remained in the study for 22 months and since the start date varied over the ICUs the total duration of the trial was 26 months. This duration might have been considerably shorter had a multisite trial with just one period of 6 months been used, provided a sufficient number of patients could have been recruited within this time frame.

6.4 Sample size calculations for dichotomous outcomes

A dichotomous outcome typically observed in clinical trials is whether a patient recovers from some disease or not. Examples in the social sciences are whether a pupil passes an exam and whether a client completes therapy. Sample size formulae for multisite trials with dichotomous outcomes are based on the odds ratio; those for the risk difference have not yet been presented in the literature. The sample size formulae are extensions to those presented in Section 3.1.3, and very similar to those for multisite trials with continuous outcomes as presented in the previous section.

6.4.1 Odds ratio

The logistic regression model is used to relate the probability p_{ij} of an outcome $y_{ij} = 1$ of subject i in cluster j to treatment condition x_{ij}:

$$p_{ij} = \frac{1}{1 + e^{-(\gamma_{00} + \gamma_{10} x_{ij} + u_{0j} + u_{1j} x_{ij})}}. \tag{6.27}$$

As in the previous section, treatment condition is coded +0.5 for the intervention and -0.5 for the control. The slope γ_{10} is the treatment effect on the logit scale, which is equal to the natural logarithm of the odds ratio as defined by (3.15). The random effects u_{0j} and u_{1j} are assumed to follow a normal distribution with zero mean and variances σ^2_{u0} and σ^2_{u1}; their covariance is denoted σ_{u01}.

The standard error of the treatment effect estimator $\hat{\gamma}_{10}$ is calculated as

$$s.e.(\hat{\gamma}_{10}) = \sqrt{\frac{4(\sigma^2 + \frac{1}{4} n_1 \sigma^2_{u1})}{n_1 n_2}}, \tag{6.28}$$

where

$$\sigma^2 = \frac{1}{2}\left(\frac{1}{p_C(1-p_C)} + \frac{1}{p_E(1-p_E)}\right). \tag{6.29}$$

p_E and p_C are the probabilities in the experimental and control conditions. The reader may note that Equation (6.28) for dichotomous outcomes is equal to Equation (6.12) for continuous outcomes once the variance component σ^2_e at the subject level is replaced by its counterpart σ^2 from Equation (6.29). This implies that results from the previous section also apply to dichotomous outcomes. Thus, power increases with increasing number of clusters and cluster size, but increasing the cluster size alone may not be sufficient to achieve a desired power level. Furthermore, power increases with increasing effect size but decreases with increasing σ^2_{u1}. The latter effect is stronger for larger cluster sizes.

Asymptotically, the test statistic $\hat{\gamma}_{10}/s.e.(\hat{\gamma}_{10})$ follows a standard normal distribution under the null hypothesis of no treatment effect. This implies that the sample size formulae for continuous outcomes as presented in the previous section can also be used for dichotomous outcomes once σ^2_e is replaced by σ^2 from Equation (6.29).

Equations (6.28) and (6.29) are based on the first order MQL estimation method, which is known to produce biased estimates of fixed and random effects. Second order PQL performs much better but, unfortunately, no closed-form expressions for $s.e.(\hat{\gamma}_{10})$ can be derived for this estimation method. A simulation study showed that the standard error for second order PQL is about 1.1 times as large as that for first order MQL (Moerbeek et al., 2001a). This implies that the sample sizes as calculated on basis of the formulae for continuous outcomes in the previous section (with σ^2_e replaced by σ^2 from

Equation (6.29)) are slightly too low and a conversion factor should be used. The correct sample sizes should be calculated from the relation

$$1.1\sqrt{\frac{4(\sigma^2+\frac{1}{4}n_1\sigma_{u1}^2)}{n_1n_2}}=\frac{\gamma_{10}}{z_{1-\alpha}+z_{1-\beta}}. \tag{6.30}$$

Constant treatment effect

If one expects the treatment effect not to vary across clusters, then the regression model (6.27) becomes

$$p_{ij}=\frac{1}{1+e^{-(\gamma_{00}+\gamma_{10}x_{ij}+u_{0j})}}. \tag{6.31}$$

The corresponding standard error of the treatment effect estimator is now equal to

$$s.e.(\hat{\gamma}_{10})=\sqrt{\frac{4\sigma^2}{n_1n_2}}, \tag{6.32}$$

with σ^2 as in Formula (6.29). Again, this equation is based on first order MQL and a conversion factor of 1.1 should be used to obtain the standard error of the treatment effect estimator for second order PQL. Equation (6.30), with $\sigma_{u1}^2=0$, can be used to calculate the power level of a given design (n_1,n_2).

6.5 An example

Ronan, Gerhart, Dollard, and Maurelli (2010) studied the effects of three treatment conditions on re-arrest of inmates housed in a county jail in rural Michigan. Twenty inmates were assigned to a no-treatment control group, 40 to a traditional self-control training group and 40 to a traditional self-control training group with behavioral rehearsal. Re-arrest was recorded every 6 months for 3 years following release from jail. A survival analysis showed significant differences between the treated and non-treated inmates. The best results were found in the traditional self-control training group with behavioral rehearsal.

Suppose a researcher wishes to conduct a similar trial to study the effects of the traditional self-control training group with behavioral rehearsal relative to the control group on the percentage of re-arrest after half a year since release from jail. The study should be designed in such a way that an effect should be detected with 80% probability in a test with a one-sided alternative hypothesis and a type I error rate $\alpha=0.05$. Power calculations are based on the odds ratio; the formulae in Section 3.1.3 show that the percentages of re-arrest in both treatment groups should be known to perform a power analysis. Of course, these values are unknown at the design stage but estimates from the trial by Ronan et al. (2010) can be used. In their study, the percentages of re-arrested after half a year were 15% in the control group and 7.5% in the self-control plus behavioral rehearsal training group.

The odds ratio is calculated from (3.15) and equal to $OR = 0.459$, with logarithm $log(OR) = -0.778$. Its corresponding standard error follows from (3.16) and is equal to $s.e.(log(OR)) = 1.055$ if as few as 20 inmates per treatment are available. From Equation (3.17) the corresponding power level is calculated to be equal to 0.173, which is clearly too low.

Including more than one jail in the trial is a strategy to improve power. Power calculations should then be based on Equation (6.30) and a prior estimate of the slope variance σ^2_{u1} is needed. An estimate of this parameter was not reported by Ronan et al. (2010) since they restricted their trial to one jail.

An educated guess of σ^2_{u1} may be obtained from the following reasoning. The parameter β_{1j} is the log odds ratio corresponding to treatment for clinic j. From (6.2) it is observed that it is a function of the average log odds ratio (denoted γ_{10}) and a random effect for jail j (denoted u_{1j}). The random effect follows a normal distribution with zero mean and variance σ^2_{u1}; thus 95% of the log odds ratios across all jails lie within the interval $(\gamma_{10} - 1.95\sigma_{u1}, \gamma_{10} + 1.95\sigma_{u1})$. The interval for the odds ratio follows from exponentiation:

$$\left(e^{\gamma_{10}-1.96\sigma_{u1}}, e^{\gamma_{10}+1.96\sigma_{u1}}\right). \tag{6.33}$$

For the trial at hand the average odds ratio is 0.459. Its range is equal to (0.247,0.854) for $\sigma^2_{u1} = 0.1$, it is equal to (0.157,1.345) for $\sigma^2_{u1} = 0.3$, and it is equal to (0.115,1.838) for $\sigma^2_{u1} = 0.5$. As is obvious, the range becomes wider if the variance σ^2_{u1} increases. The researcher should argue which range seems reasonable for the trial at hand and use the corresponding variance σ^2_{u1} to calculate the power for a given design (n_1, n_2). Assume, for instance, that $\sigma^2_{u1} = 0.1$ seems the most reasonable value of the between-jail variance. Then 15 jails with 40 inmates each are needed to achieve a power of 80% to detect a significant difference between both treatment conditions.

7

Pseudo cluster randomized trials

7.1 Introduction

Sometimes neither the cluster randomized trial (Chapter 4) nor the multisite trial (Chapter 6) is a satisfactory choice for the design of a trial. Such a dilemma arises when randomization at subject level risks control group contamination (Craven, Marsh, Debus, & Jayasinghe, 2001; Keogh-Brown et al., 2007; Plewis & Hurry, 1998; Slymen, Elder, Litrownik, Ayala, & Campbell, 2003; Rhoads, 2011), while randomization at cluster level risks selection bias as well as slow and/or incomplete recruitment.

To clarify this, consider the EASYcare trial (Melis, Van Eijken, & Borm, 2005), in which usual care by physicians for common geriatric problems of frail elderly people (e.g., falls, dementia) was compared to a nurse-led program which involved feedback by the nurse to the physician. Had randomization on patient (subject) level been used, then the physician could learn which actions and services were initiated by the nurse in the experimental arm and could use these for patients in the control condition. This would have reduced the contrast between the nurse-led program and usual care (contamination). This means that only the contaminated effect could have been estimated. The contaminated effect is smaller than the true effect and hence statistical significance for the contaminated effect implies statistical significance for the (not-estimated) true effect. However, to obtain sufficient power for the contaminated effect, the sample size would have need to be larger (Moerbeek, 2005; Slymen & Hovell, 1997).

While some authors advocate to not consider this a problem (Torgerson, 2001; Hewitt, Torgerson, & Miles, 2008), the uncontaminated effect is often considered of primary interest and hence contamination should be avoided. To achieve this in the EASYcare trial, randomization could have been done at physician (cluster) level, and this also requires a larger sample due to the clustering (Chapter 4).

Another disadvantage of cluster randomization is that it is often impossible to recruit patients before randomization of the clusters. In the EASYcare trial this was the case, because patients were recruited when they consulted their physician for a problem. Consequently, the physicians knew in advance which treatment their recruited subjects would receive and this may have influenced their choice of subjects (selection bias), which leads to differences between the treatment groups at baseline (Hahn et al., 2005; Jordhoy, Fayers, Ahlner-Elmqvist, & Kaasa, 2002; Puffer, Torgerson, & Watson, 2003).

Moreover, advance knowledge of treatment allocation may also influence the rate of recruitment. Typically physicians participate in such a trial because this gives them the opportunity to learn new skills and/or offer their patients improved care. Physicians may then lose interest if they can only apply the control treatment to patients and this may result in recruiting fewer subjects and producing data of poorer quality (Klar & Donner, 2001; Moore, Summerbell, Vail, Greenwood, & Adamson, 2001; Farrin, Russell, Torgerson, & Underwood, 2005).

In situations like the EASYcare trial, where neither the cluster randomized trial nor the multisite trial is a satisfactory design choice, a pseudo cluster randomized trial (Moerbeek & Teerenstra, 2010; G. F. Borm, Melis, Teerenstra, & Peer, 2005; Teerenstra, Melis, Peer, & Borm, 2006; Melis et al., 2005; Melis, Teerenstra, Olde Rikkert, & Borm, 2008) can be considered.

Pseudo cluster randomization (Moerbeek & Teerenstra, 2010; G. F. Borm, Melis, et al., 2005; Teerenstra et al., 2006; Melis et al., 2005, 2008) combines randomization at subject level with randomization at cluster level in the following way (see Figure 1.4): first the clusters are randomized into two types, E and C, and the results of this randomization are not revealed. Then, within each cluster of type E, subjects are randomized in majority to the experimental condition e and, therefore, the minority receives the control condition. In the clusters of type C, the situation is reversed, so that the majority of subjects are randomized to the control condition c.

Pseudo cluster randomization is in-between randomization at subject level (as in the multisite trial) and randomization at cluster level (as in the cluster randomized trial) and encompasses both types of randomization as extreme cases. The cluster randomized design occurs if in clusters of type E (C) not only a majority of subjects, but all of the subjects are randomized to e (c). The multisite trial arises when in each cluster of type E (C) half of the subjects are randomized to e (c).

How does pseudo cluster randomization help to solve the dilemma of the multisite trial with contamination on the one hand versus the cluster randomized trial with selection bias and/or recruitment issues on the other hand (Melis, Teerenstra, Olde Rikkert, & Borm, 2011)? Consider the EASYcare trial again. Due to the two-step randomization, the physicians do not know in which type of cluster they are, nor do they know in advance what treatment the next patient will receive. This reduces the chance of selection bias. Moreover, no longer half of the physicians is left with only patients on the uninteresting treatment (as in cluster randomization). In fact, all physicians will receive patients on the interesting treatment, which may improve recruitment. Finally, in clusters of type E, most patients are on condition e and only a few are on condition c. This means that the contamination of e by c is small. On the other hand, the contamination of c by e may be substantial, but that only affects the few patients on c. Similar considerations apply to clusters of type C. Thus the impact of contamination may be smaller in pseudo cluster randomization than in individual randomization.

Upon closer examination, pseudo cluster randomization will reduce selection bias and recruitment issues to the extent that the predictability of the treatment allocation sequence is reduced. Obviously, the physician can never be sure about which condition will be allocated next, so predictability will always be less than in a cluster randomized design. Nevertheless, the physician may guess how likely condition e will be allocated in his or her cluster. By not revealing the type of cluster the physician is randomized to, by keeping the cluster size small, and by making the imbalance in allocation not too extreme (e.g., $f = 0.8$), this effect can be reduced, so that it would be hard for a physician to guess which cluster group he or she is allocated to.

In fact, the physicians in the EASYcare trial in large majority thought that 1:1 randomization was used (Melis et al., 2008), while actually $f = 0.8$ was used. Possibly, pseudo cluster randomization also increased recruitment as the majority of physicians had a strong preference for their patients to be randomized to intervention and reported to have recruited fewer patients if all their patients would be on usual care (Melis et al., 2008).

Pseudo cluster randomization will reduce contamination if the degree of contamination is proportional to exposure in the following sense: control subjects in a cluster will be less contaminated (or fewer control subjects will be contaminated) if there are fewer subjects on the experimental condition in that cluster. This would be an unreasonable assumption if one single patient on the experimental condition will lead to complete contamination of all other patients. An example of this would be if the experimental condition consists of a flyer containing a few easy to learn skills and/or easy to obtain services suitable for all patients (i.e., non-tailored). However, the assumption that contamination is proportional to exposure is reasonable if the dissemination of elements of the experimental condition to the control group is a gradual process that depends on the number of subjects on the experimental condition in each cluster.

In the EASYcare trial, the intervention was delivered by a nurse during visits at the homes of patients, but the physician could learn elements the nurse applied during these visits from the feedback meetings with the nurse. However, the intervention was not fast and easy to copy to control patients, because the intervention by each nurse was highly tailored to the patient and required specialist geriatric nursing skills and knowledge. Finally, the opportunities to learn were limited because the number of patients in the experimental condition was limited in the clusters where the majority of patients received intervention (the type E clusters). For contamination the other way round, which is mostly interpreted as non-compliance, similar considerations apply in principle.

The above considerations delineate in which settings pseudo cluster randomization may be useful: trials in which the clusters are small and where contamination is not immediate but takes several subjects in the experimental condition before contamination of subjects in the control condition reaches a substantial level. Examples of such trials are complex interventions in health care (e.g., multi-faceted and/or multi-disciplinary and/or tailored-to-the patient interventions in hospital wards, centers of expertise, or in small populations within general practices such as frail elderly).

In comparison with other methods to address contamination, selection bias and recruitment issues, pseudo cluster was considered a good compromise in the design of the EASYcare trial (Melis et al., 2011). As there is no one-size-fits-all approach to designing a trial, it is recommended to consider in the design phase a range of options to address recruitment (Flynn, Whitley, & Peters, 2002; Williamson et al., 2007; Hoddinott, Britten, Harrild, & Godden, 2007; Dyas, Apekey, Tilling, & Siriwardena, 2009), selection bias (Kerry, Cappuccio, Emmett, Plange-Rhule, & Eastwood, 2005) and other sources of biases (Giraudeau & Ravaud, 2009).

Like the multisite trial and the cluster randomized trial, the pseudo cluster randomized trial applies to nested data and this clustering has to be accounted for in the design and analysis. Therefore, Section 7.2 will provide a multilevel model for the pseudo cluster randomized trial, while Sections 7.3 and 7.4 provide the sample size calculations for continuous and binary outcomes, respectively. Section 7.5 will show how to consider different designs and calculate their sample size using the EASYcare trial as an example.

7.2 Multilevel model

The outcome y_{ij} of subject i $(= 1, \ldots, n_1)$ in cluster j $(= 1, \ldots, n_2)$ in a pseudo cluster randomized trial can be formulated:

$$y_{ij} = \gamma_0 + \gamma_1 x_{ij} + u_j + e_{ij}, \tag{7.1}$$

which is similar to the formulation for a cluster randomized trial, except that the treatment indicator x_{ij}, taking values 0 and 1, now varies over subjects in a cluster. In the above formulation, the average effect of the experimental condition over the control condition is estimated, where the average is over all clusters. In particular no difference in the effect is assumed (estimated) between clusters of type E and C or between clusters within the same type. If sufficient clusters are available, corresponding interaction effects could be estimated as well by adding appropriate interaction terms.

Assume all clusters are equally randomized to both types and in each cluster the majority $0.5 < f < 1$ receives the condition corresponding to the cluster type. Then in each of the $n_2/2$ clusters of type E with cluster size n_1, $n_1 f$ subjects receive condition e (Ee subjects) and $n_1(1-f)$ receive the control condition c (Ec subjects). Similarly, in each of the $n_2/2$ clusters of type C, $n_1 f$ subjects receive the control condition (Cc subjects) and $n_1(1-f)$

subjects receive the experimental condition (Ce). Clearly, $f = 1$ would then correspond to the cluster randomized trial and $f = 0.5$ to the multisite trial (without cluster by treatment interaction). For a pseudo cluster randomized trial, $f = 0.8$ seems to be a good choice in general (G. F. Borm, Melis, et al., 2005), although this choice should be made on a trial-by-trial basis.

The effect of condition e can be estimated from both the Ee subjects and the Ce subjects. Let $\bar{y}_j^{Ee}$ and $\bar{y}_j^{Ce}$ be the average over the Ee subjects in cluster j of type E or C, respectively. As there are fewer subjects on condition e in type C clusters than in type E clusters, the estimates $\bar{y}_j^{Ce}$ will be less precise than the estimates $\bar{y}_j^{Ee}$. Therefore, it may be more efficient to weight the $\bar{y}_j^{Ce}$ and estimate the effect on condition e as:

$$\bar{y}_e = \frac{\frac{1}{n_2/2}\sum f\bar{y}_j^{Ee} + \frac{1}{n_2/2}\sum w(1-f)\bar{y}_j^{Ce}}{f + w(1-f)}$$

and similarly, the effect on condition c can be estimated as the weighted average:

$$\bar{y}_c = \frac{\frac{1}{n_2/2}\sum f\bar{y}_j^{Cc} + \frac{1}{n_2/2}\sum w(1-f)\bar{y}_j^{Ec}}{f + w(1-f)},$$

so that

$$\bar{y}_e - \bar{y}_c = \frac{1}{n_2/2}\sum_{j=1}^{n_2/2}\left[\alpha\bar{y}_j^{Ee} - \beta\bar{y}_j^{Ec}\right] + \frac{1}{n_2/2}\sum_{n_2/2+1}^{n_2}\left[\beta\bar{y}_j^{Ce} - \alpha\bar{y}_j^{Cc}\right] \tag{7.2}$$

where $\alpha = f/[f + w(1-f)]$ and $\beta = w(1-f)/[f + w(1-f)]$ weight the estimates. In each cluster, both conditions are present and hence part of the between-cluster variance is removed. To see this, consider

$$\begin{aligned} &var(\alpha\bar{y}_j^{Ee} - \beta\bar{y}_j^{Ec}) = \\ &var\left(\alpha\mu_{Ee} - \beta\mu_{Ec} + (\alpha-\beta)u_j + \alpha\frac{\sum_{i=1}^{n_1 f} e_{ij}}{n_1 f} - \beta\frac{\sum_{i=n_1 f+1}^{n_1} e_{ij}}{n_1(1-f)}\right) \\ &= (\alpha-\beta)^2\sigma_u^2 + \alpha^2\frac{\sigma_e^2}{n_1 f} + \beta^2\frac{\sigma_e^2}{n_1(1-f)}, \end{aligned} \tag{7.3}$$

and observe that $(\alpha-\beta)^2 < 1$ if $0.5 < f < 1$.

To minimize the variance of the estimator, weights should be chosen as

$$w = \frac{1 + 2fn_1(\sigma_u^2/\sigma_e^2)}{1 + 2(1-f)n_1(\sigma_u^2/\sigma_e^2)} = \frac{1 + 2fn_1\rho/(1-\rho)}{1 + 2(1-f)n_1\rho/(1-\rho)} \tag{7.4}$$

which are functions of $\sigma_u^2/(\sigma_e^2/n_1) = n_1\rho/(1-\rho)$, that is, the ratio of the between- and within-variance of the cluster *average*. Other objectives of weighting are discussed in G. F. Borm, Melis, et al. (2005).

Then the standard error of the treatment effect estimator $\bar{y}_e - \bar{y}_c$ is

$$s.e.(\bar{y}_e - \bar{y}_c) = \sqrt{\frac{4\sigma^2}{n_1 n_2} \frac{1 + (n_1 - 1)\rho}{1 + 4f(1-f)n_1\rho/(1-\rho)}}; \tag{7.5}$$

see G. F. Borm, Melis, et al. (2005).

7.3 Sample size calculations for continuous outcomes

Pseudo cluster randomized trial data as in Equation (7.1) can be analyzed using the same multilevel model as that of a cluster randomized trial before (i.e., a random effect for cluster and a fixed effect for the treatment indicator). The only difference is that the treatment indicator in the pseudo cluster randomized design now varies both at cluster and individual level. Alternatively, the same multilevel model as that of a multisite trial can be taken, but without treatment by cluster interaction.

As the multilevel estimator of Equation (7.1) is equivalent with the estimator $\bar{y}_e - \bar{y}_c$ for a fixed cluster size n_1 (Teerenstra, Moerbeek, Melis, & Borm, 2007), the power and sample size follow from (3.20) and (7.5). On the basis of that, the factors that influence power, the design effect, and the sample size calculation will be discussed below.

7.3.1 Factors that influence power

The results that are presented in this subsection assume the following configuration. All clusters have equal size of n_1 subjects and the majority fraction $0.5 \leq f \leq 1$ is equal for all clusters. Half of the clusters ($n_2/2$ clusters) are of type E with $n_1 f$ subjects receiving condition e and $n_1(1-f)$ receiving the control condition c (Ec subjects). Similarly, the remaining half of the clusters ($n_2/2$ clusters) are of type C with $n_1 f$ subjects receiving the control condition and $n_1(1-f)$ subjects receiving the experimental condition.

Number of clusters. The number of clusters n_2 only occurs in the denominator of the standard error; see (7.5). Therefore, when the number of clusters n_2 becomes very large, the standard error (7.5) will approach zero. Consequently, power can increase arbitrarily close to 100% whenever the number of clusters is large enough. This is a similar finding as for cluster randomized trials and multisite trials (see Section 4.3.1, and Section 6.3.1).

Cluster size. The cluster size n_1 occurs once in the numerator and in two factors of the denominator of the standard error unless $f = 1$; see (7.5). Therefore, as the cluster size becomes very large, the standard error approaches zero (unless $f = 1$, which corresponds to the cluster randomized trial) and power can reach 100% when the cluster size is taken large enough. This is a feature similar to the multisite trial with no treatment by cluster interaction (see Equation (6.13) with $\rho_1 = 0$).

This can be explained as follows: in each cluster there are subjects in the experimental condition and in the control condition. If the cluster size is large, then the estimates in the experimental condition and in the control condition are equally weighted: for large n_1, the weight $w \to f/(1-f)$, see Equation (7.4), and $\alpha = \beta = 1/2$ in Equation (7.3). In other words, the part of the variance due to variation of clusters is removed if the cluster size is large. Moreover, the precision of this difference of means depends essentially on the number of subjects in the cluster (and the degree of imbalance of the allocation ratio f). Hence with larger clusters, more precision is obtained (unless $f = 1$).

Intracluster correlation. In general, the standard error will first slightly increase and then decrease as the ICC increases (i.e., initially power slightly decreases and then increases). This increase is more pronounced if f is closer to 1 and absent if $f = 0.5$. This is due to the interplay of the numerator $1 + (n_1 - 1)\rho$ and the denominator $1 + 4f(1-f)n_1\rho/(1-\rho)$ in the standard error (7.5). The numerator describes the usual clustering effect of the cluster randomized trial: if ρ increases, this clustering effect increases the numerator. The denominator describes the effect of removing between-cluster variance.

The larger the ICC, the larger the between-cluster variance and therefore the larger the denominator. If $f = 0.5$, then $f(1-f)$ will be maximal and the denominator is larger than the numerator even for small values of ρ: the standard error will decrease as ρ increases. If f moves to 1, then $f(1-f)$ is smaller and the denominator will initially be smaller than the numerator for small values of ρ. Actually, this says that the imbalance in allocation overrules the removing of between-cluster variance when the between-cluster variance (i.e., the ICC) is small.

Majority fraction f. The closer f is to 0.5 the less unbalanced the allocation is within clusters and hence the smaller the standard error and the larger the power. As discussed above, f also has an influence via the ICC. However, f has an additional influence: the standardized effect δ actually depends on f. This is easily seen by first considering the extremes. For $f = 1$, which corresponds to the cluster randomized trial, the contamination will be (ideally) absent and the treatment effect is equal to $\Delta_1 = (\mu_e - \mu_c)$, based on the difference between the expected outcomes in the experimental and control conditions *in absence of contamination.* For $f = 0.5$, which corresponds to the multisite trial, cross-exposure of the conditions in each cluster will be maximal, and the contamination of the effect is expected to be maximal. If f is between 0.5 and 1, the expected outcome of condition e in the subset Ee will be $\mu_{Ee} = \mu_e - c_{major}\Delta_1$, where $0 \leq c_{major} \leq 1$ is the one-sided relative contamination of the majority (the Ee-patients) by the minority (the Ec-patients) in each cluster of type E.

In Ce, the expected outcome is $\mu_{Ce} = \mu_e - c_{minor}\Delta_1$ where $0 \leq c_{minor} \leq 1$ is the one-sided relative contamination of the minority (the Ce-patients) by the majority (the Cc-patients) in each cluster of type C. Similarly, the expected outcomes in treatment groups Cc and Ec are $\mu_{Cc} = \mu_c + c_{major}\Delta_1$ and $\mu_{Ec} = \mu_c + c_{minor}\Delta_1$. The contaminated treatment effect, as estimated by $\bar{y}_e - \bar{y}_c$, has then as expected value

$$\Delta = \Delta_1 - \Delta_1 \frac{2f c_{major} + 2w(1-f)c_{minor}}{f + w(1-f)}. \tag{7.6}$$

Dividing by the standard deviation σ results in the expected standardized treatment effect δ_f which depends on f as follows:

$$\delta = \delta_1 - \delta_1 \frac{2f c_{major} + 2w(1-f)c_{minor}}{f + w(1-f)}. \tag{7.7}$$

Reasonably, the following would be expected. First, the contamination of a minority by a majority would larger than vice versa ($c_{minor} \geq c_{majority}$). Second, when f moves from 0.5 to 1, the contamination of the majority c_{major} will decrease to its minimum (since the cause of this contamination, the minority, decreases). Third, when f moves from 0.5 to 1, the contamination of the minority c_{minor} will increase to its maximum (since the cause of this contamination, the majority, increases in size). However, the above argumentation has to be considered on a trial-by-trial basis, since this depends on the characteristics of subjects, the cluster, the intervention and their interplay.

7.3.2 Design effect

From the standard error (7.5), the design effect of pseudo cluster randomization compared to individual randomization follows as

$$\text{design effect} = \frac{1 + (n_1 - 1)\rho}{1 + 4f(1-f)n_1\rho/(1-\rho)}; \tag{7.8}$$

see Teerenstra et al. (2006). For $f = 1$ this indeed results in the usual design effect of cluster randomization. For $f = 0.5$ (the multisite trial), the design effect reduces to $1 - \rho$. This may seem surprising at first sight (if one expects 1 because of individual randomization), but on second thought this is understandable. The design effect is compared to *unstratified* individual randomization of subjects. In the multisite trial, the randomization is by cluster and hence the between-cluster variance is removed. Therefore, the variance in the multisite trial is only the within-cluster variance. When clustering is absent, that is, if there are no clusters ($n_1 = 1$) or no correlation within the clusters ($\rho = 0$), the design effect equals 1, as expected.

7.3.3 Sample size formulae for fixed cluster size or fixed number of clusters

The sample size for a fixed cluster size n_1 and equal number of clusters in each group $n_2/2$ satisfies the following relation

$$\begin{aligned} n_2 n_1 &= 4\frac{\sigma_e^2 + n_1\sigma_u^2}{1+4f(1-f)n_1(\sigma_u^2/\sigma_e^2)} \cdot \left(\frac{z_{1_\alpha/2} + z_{1-\beta}}{\Delta}\right)^2 \\ &= 4\frac{1+(n_1-1)\rho}{1+4f(1-f)n_1\rho/(1-\rho)} \cdot \left(\frac{z_{1_\alpha/2} + z_{1-\beta}}{\delta}\right)^2 \end{aligned} \qquad (7.9)$$

from which the number of clusters n_2 can be calculated given the cluster size by dividing the right sides of the equation by n_1. The reverse (determining the cluster size given the number of clusters n_2^*) involves solving a quadratic equation. A more pragmatic approach for this could be to list the number of clusters needed for various cluster sizes and looking whether there is a cluster size that corresponds to the preset number of clusters n_2^*.

Using (7.9) and (7.7), the relative efficiency of pseudo cluster randomization can be compared to that of cluster randomization and individual randomization stratified by cluster (without treatment by cluster interaction). Figure 7.1 shows the relative efficiency of pseudo cluster randomization and individual randomization compared to that of cluster randomization (which is set to 1). The scale on the horizontal axes is the ratio of the between- and within-variance of the cluster average: $\sigma_u^2/(\sigma_e^2/n_1) = n_1\rho/(1-\rho)$ which is abbreviated as $n_1 q$ where $q = \sigma_u^2/\sigma_e^2$. The contamination of the minority by a majority in a cluster is set to 0.2, while the contamination of the majority by the minority is either absent ($c_{major} = 0$) or half that of the contamination of the minority ($c_{major} = 0.1$).

In general, cluster randomization is most efficient for small values of $n_1 q$ and individual randomization is better for large values of $n_1 q$. In the range between, pseudo cluster generally outperforms both, while its efficiency is close to that of individual randomization for large values of $n_1 q$, especially if in each cluster the contamination of the majority (by the minority) is small compared to the contamination of the minority (by the majority). In conclusion, the main reasons for applying pseudo cluster randomization are methodological (to reduce selection bias compared to a cluster randomized design and/or to reduce contamination compared to an individually randomized design) or practical (to improve recruitment compared to a cluster randomized design). Nevertheless, pseudo cluster may also be more efficient than either cluster or individual randomization which can be seen in the example of the Dutch EASYcare study; see Section 7.5.

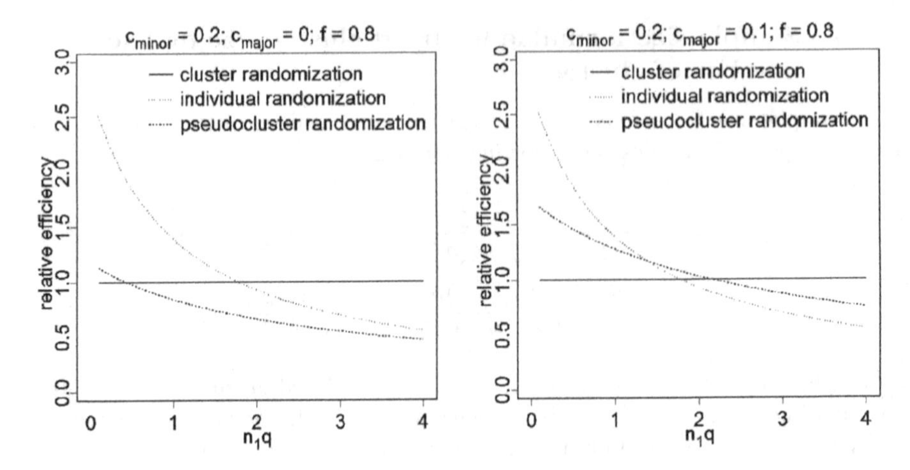

Figure 7.1: Relative efficiency of pseudo cluster and individual randomization compared to that of cluster randomization as a function of the ratio of the between- and within-variance of cluster average n_1q.

7.4 Sample size calculations for binary outcomes

In case of a binary outcome, for example, success (coded 1 =yes and 0 =no), the effect can be estimated as the difference between the proportion success in the experimental condition and that in the control condition. Under assumptions as explained below, the sample size calculation for this effect measure follows the same lines as that for continuous outcomes after obvious modifications. The first modification is that means in the subsection *Majority fraction* f in Section 7.3.1, should be replaced by proportions. Thus, μ_e and μ_c are the expected proportions of success under the ideal situation of no contamination, μ_{Ee} and μ_{Ce} are the expected proportions of successes in the experimental condition in clusters of type E and C, and likewise μ_{Ec} and μ_{Cc} for the control condition. The next modification is that estimators $\bar{y}_j^{Ee}$, $\bar{y}_j^{Ec}$, $\bar{y}_j^{Cc}$, $\bar{y}_j^{Ce}$ of Section 7.2 are now averages of a binary outcome and hence estimate proportions: μ_{Ee}, μ_{Ec}, μ_{Cc} and μ_{Ce} respectively. In particular, the estimator $\bar{y}_e - \bar{y}_c$ of equation (7.2) estimates the contaminated difference in proportion success as expressed by Equation (7.6).

Model. Assume instead of the linear multilevel model (7.1) a hierarchical binomial model (Hendriks et al., 2005):

1. *Level 1 (subject level) model.* Independently of each other, the subjects in cluster j have outcome 'success' with probability p_j when in the experimental condition and with reduced probability $p' = p - r$ when in the control condition.
2. *Level 2 (cluster level) model.* The success rate p varies over the

clusters according to a $[0, 1]$-valued distribution D with mean M and variance Σ^2, for example, D could be a beta-distribution. $M = \mu_{Ee}$ and $M - r = \mu_{Ec}$ in clusters of type E. Likewise, $M = \mu_{Ce}$ and $M - r = \mu_{Cc}$ in clusters of type C.

In particular, the results that follow below hold approximately with the approximation being accurate if the probability that p is below r is very small (see the Appendix of Hendriks et al. (2005)).

For the ease of notation abbreviate $P_e = \bar{y}_j^{Ee}$ and $P_c = \bar{y}_j^{Ec}$, then the counter part of Equation (7.3), is

$$\begin{aligned} var(\alpha\bar{y}_j^{Ee} - \beta\bar{y}_j^{Ec}) &= var(\alpha P_e - \beta P_c) \\ &= var\left[E\left(\alpha P_e - \beta P_c|p\right)\right] + E\left[var\left(\alpha P_e - \beta P_c\right)|p\right)\right]. \end{aligned}$$

The first term is worked out as follows. In cluster j of type E, $p = p_j$ and $\bar{y}_j^{Ee} = \frac{1}{m}\sum_{i=1}^{m} x_{ij}$ is distributed with expected value p and variance $p(1-p)/m$ and likewise $\bar{y}_j^{Ec}$ has expected value $p - r$ and variance $(p - r)(1 - [p - r])/k$, where m and k abbreviate $n_1 f$ and $n_1(1 - f)$ respectively. Then the first term

$$var\left[E\left(\alpha P_e - \beta P_c|p\right)\right] = var\left[\alpha p - \beta(p - r)\right] = (\alpha - \beta)^2\Sigma^2$$

describes the between-cluster variance. The second term is calculated using $E\left[(\cdot)^2\right] = var(\cdot) + \left[E(\cdot)\right]^2$:

$$\begin{aligned} E\left[var\left(\alpha P_e - \beta P_c\right)|p\right)\right] &= E\left[\frac{\alpha^2}{m}\left(p - p^2\right) + \frac{\beta^2}{k}\left((p - r) - (p - r)^2\right)\right] \\ &= \frac{\alpha^2}{m}\left[M - (\Sigma^2 + M^2)\right] + \frac{\beta^2}{k}\left[(M - r) - (\Sigma^2 + (M - r)^2)\right] \\ &= \frac{\alpha^2}{m}\left[M(1 - M) - \Sigma^2\right] + \frac{\beta^2}{k}\left[(M - r)(1 - [M - r]) - \Sigma^2\right] \end{aligned}$$

which describes the within-cluster variance. Thus,

$$\begin{aligned} &var(\alpha\bar{y}_j^{Ee} - \beta\bar{y}_j^{Ec}) \\ &= (\alpha - \beta)^2\Sigma^2 + \alpha^2\frac{\left[M(1 - M) - \Sigma^2\right]}{n_1 f} - \beta^2\frac{\left[(M - r)(1 - [M - r]) - \Sigma^2\right]}{n_1(1 - f)} \end{aligned} \tag{7.10}$$

is the counterpart of (7.3) for binary outcomes.

Note that all of μ_{Ee}, μ_{Ce}, μ_{Cc}, μ_{Ec} lie between μ_c and μ_e. If μ_e and μ_c are within 0.2 and 0.8, M and $M - r$ will be close to $[\mu_e(1 - \mu_e) + \mu_c(1 - \mu_c)]/2$. Approximate sample size calculations can then be performed as in the continuous case (Section 7.3) using the ICC ρ and identifying $\sigma^2 = [\mu_e(1 - \mu_e) + \mu_c(1 - \mu_c)]/2$. In fact, the analogy between the linear multilevel model and the hierarchical binomial model is $\sigma_e^2 \leftrightarrow [\mu_e(1 - \mu_e) + \mu_c(1 - \mu_c)]/2 - \Sigma^2$, $\sigma_u^2 \leftrightarrow \Sigma^2$, and $\rho \leftrightarrow \Sigma^2/([\mu_e(1 - \mu_e) + \mu_c(1 - \mu_c)]/2)$.

If μ_e or μ_c is close to 1 or 0, the recommendation is to base the sample size on a simulation study.

7.5 An example

The Dutch EASYcare trial was an observer-blinded trial designed to compare an intermediate care program for elderly people with geriatric problems to the traditional care as provided by physicians. The primary endpoint was GARS-3, a functional performance measure for daily life activities that ranges from 18 to 54.

A pilot study on the target population showed that the GARS had a standard deviation $\sigma = 8.5$ (and a mean $\mu = 35$). It was expected that the intermediate care program would improve a patient's capacity to perform daily life activities and result in a higher average score on the GARS that could amount to $\delta = 4.5$ in the *absence* of contamination. At a two-sided significance level of $\alpha = 0.05$, the objective was to achieve a power of 80% ($\beta = 0.2$).

The impact of each randomization scenario on the total sample size is considered below. If *cluster randomization* had been chosen for the EASY-care trial, contamination would have been absent, and the expected treatment difference would have been $\delta = 4.5$. However, clustering introduces a positive correlation between the patients in a cluster. This is expressed by the intracluster correlation coefficient, which was anticipated to be $\rho = 0.05$ in this trial. Following the usual procedure, the uncorrected sample size per each treatment arm would be calculated using (3.5), which would yield 56. This value would then be corrected by multiplying it with the design factor $1 + (n_1 - 1)\rho$. As the maximum cluster size expected was $n = 10$, the result would be a total sample size of 164.

If *individual randomization* had been used for the EASYcare study, contamination would have been substantial, and it would be anticipated to dilute the expected treatment difference to $\delta = 3.5$. However, no clustering would take place and using (3.5), the sample size of each treatment arm would be 93, that is, the total sample size would be 186.

In fact, *pseudo cluster randomization* was used in the EASYcare trial. Thus, the rate of contamination was expected to decrease compared to the individually randomized situation ($\delta = 3.5$), although not as far as the uncontaminated level ($\delta = 4.5$) of a cluster randomized setting. An expected treatment difference of $\delta = 4$ was deemed realistic. Following the procedure outlined above, the total sample size was calculated to be 154 using (7.9) with $n = 10$, $\rho = 0.05$, $f = 0.8$. The above calculations showed that pseudo cluster randomization had an additional advantage in the Dutch EASYcare trial. The sample size was smaller than those required for cluster randomization and individual randomization. This is not uncommon as illustrated in Figure 7.1.

8

Individually randomized group treatment trials

8.1 Introduction

The designs discussed in the previous chapters share the feature that nesting of individuals within clusters was established prior to randomization to treatment conditions. Examples are trials with pupils within schools, family members within families and patients within clinics. Randomization is done at the cluster level in case of a cluster randomized trial (Chapters 4 and 5) and at the individual level in case of a multisite trial (Chapter 6). With pseudo cluster randomized trials, a two-stage randomization procedure is used as explained in Chapter 7. In either case, it is assumed that randomization does not influence group membership. In other words, subjects are assumed not to change groups due to their assignment to a particular treatment.

Clustering may also occur when randomization to treatment conditions is done at the individual level but treatment is offered in groups, or when clients are randomly assigned to health professionals and multiple subjects are treated by the same health professional. In such cases, the nesting of subjects within groups or within health professionals is established once randomization is done. Following S. P. Pals et al. (2008) such trials are labeled as individually randomized group treatment trials to stress the feature that randomization is done by individual but treatment is (partially) offered in groups.

Consider as an example a trial that aims to improve the wellbeing of caregivers of stroke patients (Van Den Heuvel, De Witte, Nooyen-Haazen, Sanderman, & Meyboom-De Jong, 2000). Caregivers are randomly assigned to a group program, a home visit program or a control condition. The group program consists of sessions in groups of caregivers who are supervised by a health education nurse. The home visit program consists of visits by a district nurse. In the first condition, caregivers are nested within session groups and the outcomes on wellbeing scores of caregivers in the same group may be correlated due to mutual influence among caregivers in the same session group and the effects of the nurse. In the second condition, outcomes of caregivers who are visited by the same nurse may be correlated due to the skills, experience, motivation, and enthusiasm of the nurse. Uncorrelated data may be expected in the control condition.

Another example is a trial on the treatment of social phobia (Otte et al., 2000). Again randomization was done at the subject level. The experimental condition is cognitive-behavioral group therapy and the control condition is pharmacotherapy supplemented by weekly monitoring by a psychiatrist. Correlation of outcome variables may be expected in both conditions. In the first condition, such correlation may arise from mutual influence among patients in the same treatment session groups; in the second case, it may be due to psychiatrist effects.

These two trials are just a selection among the many published trials where clustering was induced by randomization. Clustering may occur in all treatment conditions, in just a few of them, or in only one. In the last two cases, the trial results in data with partial clustering. It may be clear that (partial) clustering must be accounted for in sample size calculations and data analysis. In the late 1970s it was already acknowledged that ignoring therapist effects in psychotherapy research leads to inflated type I error rates for tests on treatment effect (Martindale, 1978).

Reviews of studies of individually randomized group treatment trials show that this issue is too often ignored in both data analysis and sample size calculations of individually randomized trials (S. P. Pals et al., 2008; Crits-Christoph & Mintz, 1991). This may be explained by the fact that many researchers in social and biomedical sciences do not acknowledge that clustering may occur in individually randomized trials or they underestimate the effects of clustering on inferences with respect to the treatment effect. In addition, they must be familiar with multilevel models to correctly analyze the data and, as will be shown in the next section, the multilevel model for trials with partial nesting requires an extension to the random part of the multilevel model for cluster randomized trials that may not be obvious at first sight. Furthermore, power and sample sizes for trials with partial clustering have only been studied recently (Roberts & Roberts, 2005; Moerbeek & Wong, 2008; Heo, Litwin, Blackstock, Kim, & Arnsten, in press).

The appropriate statistical model for individually randomized group treatment trials is the multilevel model. In this chapter a restriction is made to trials with two treatment conditions and at most two levels of nesting (see Heo et al. (in press) for up to three levels of nesting). The multilevel model formulated in Section 4.2 can easily be used in case clustering occurs in both treatment conditions and may be extended to account for heteroscedasticity among treatment arms.

Models for trials with clustering in just one treatment arm were introduced in 2005 (Lee & Thompson, 2005; Roberts & Roberts, 2005). Prior to 2005, various models for the analysis of partially clustered data were proposed; see Aydin, Leite, and Algina (2014), Baldwin, Bauer, Stice, and Rohde (2013), Bauer, Sterba, and Hallfors (2008) and Korendijk (2012) for an overview and comparison of such models.

One approach is to pretend no observations are clustered and apply a traditional regression model. Another approach is to pretend all observations are clustered by assuming the n_c subjects in the control condition are either clustered in one cluster of size n_c, or in n_c clusters of size 1. In general, such approaches result in conclusions with respect to the treatment effect that differ from those obtained with the appropriate multilevel model that accounts for clustering in the one treatment arm but not in the other.

In case clustering occurs in both treatment conditions, the sample size formulae for cluster randomized trials as presented in Chapter 4 may be used. These formulae may be extended to account for heteroscedasticity, and different cluster sizes and intraclass correlation coefficients in both conditions. Such extensions are further discussed in Section 8.3.1. The calculation of sample sizes for trials with partial nesting cannot be calculated on basis of sample size formulae for simple or cluster randomized trials since these ignore clustering altogether or suppose all subjects are clustered in groups, irrespective of whether they are in the control or experimental condition. Moreover, three sample sizes must be considered in such trials: the number of clusters and the cluster size in the group treatment condition and the number of subjects in the control condition.

As in the previous chapters, the multilevel model is introduced first and then the focus is on sample size calculations. An example illustrates the application of the sample size formulae in a practical setting. In the equations that follow, a statistical test with a one-sided alternative hypothesis is assumed. For a two-sided alternative hypothesis, the type I error rate α needs to be replaced by $\alpha/2$.

8.2 Multilevel model

8.2.1 Clustering in both treatment arms

With cluster randomized trials, complete clusters are randomly allocated to treatment conditions and one would expect cluster sizes, intraclass correlation coefficients and total outcome variance not to differ over treatment arms. This does not necessarily hold with individually randomized group treatment trials, for instance when patients in both treatment arms are treated by different types of health professionals. Roberts (1999) compared patients who were treated by their own doctors to those treated by a deputizing doctor on two outcome variables: patients' satisfaction and proportion of patients receiving a prescription. For both outcomes, between-doctor variance differed over the two types of doctors. This was also the case in a trial that compared patients' satisfaction of nurse practitioners and general practitioners (Roberts & Roberts, 2005). This trial demonstrates that cluster sizes may vary over treatment conditions in an individually randomized group treatment trail:

the 20 nurse practitioners saw on average 30 patients, while the 71 general practitioners saw on average 8 patients.

The multilevel model (4.3) for cluster randomized models can be extended to accommodate different between- and within-cluster variation over the treatment arms:

$$y_{ij} = \gamma_0 + \gamma_1 x_j + u_j x_j + \tilde{u}_j(1 - x_j) + e_{ij} x_j + \tilde{e}_{ij}(1 - x_j). \qquad (8.1)$$

As previously, y_{ij} and x_j are the continuous outcome and dichotomous treatment indicators, respectively. Given coding $x_j = 0$ for the control condition and $x_j = 1$ for the experimental condition, the fixed effect γ_0 is the mean outcome in the control condition and the fixed effect γ_1 is the difference in outcomes between both treatment conditions. In other words, γ_1 is the unstandardized treatment effect. As with cluster randomized trials, treatment condition x_j has subscript j but not i since all subjects in the same cluster receive the same treatment condition. The random effects at the cluster and individual levels for those in the control condition are indicated with a tilde to distinguish them from those in the experimental condition. As in previous chapters, they are assumed to be normally distributed with zero mean and a tilde is used to denote the variances in the experimental condition. Thus, σ_e^2 and σ_u^2 are the variances at the individual and cluster levels for those in the experimental condition and $\tilde{\sigma}_e^2$ and $\tilde{\sigma}_u^2$ are the corresponding variances in the control condition.

It should be noted that this model differs from the multilevel model (4.3) for cluster randomized models by its inclusion of the treatment indicator x_j in the random part of the model. The model for those in the experimental condition is $y_{ij} = \gamma_0 + \gamma_1 + u_j + e_{ij}$ and for those in the control condition it is $y_{ij} = \gamma_0 + \tilde{u}_j + \tilde{e}_{ij}$. The intraclass correlation coefficients in the two treatment arms are defined as $\rho = \sigma_u^2/(\sigma_u^2 + \sigma_e^2)$ and $\tilde{\rho} = \tilde{\sigma}_u^2/(\tilde{\sigma}_u^2 + \tilde{\sigma}_e^2)$. The total variance is allowed to vary over the treatment conditions, that is, it is not necessarily assumed $\sigma_e^2 + \sigma_u^2 = \tilde{\sigma}_e^2 + \tilde{\sigma}_u^2$. In other words: the model allows for heteroscadasticity.

Depending on the cluster sizes and the sizes of the intraclass correlation coefficients in both conditions, ignoring differential cluster size and intraclass correlation may result in an under- or overestimation of the standard error of treatment effect estimator. Assume both treatment arms are of equal size and have equal total variance. The ratio of the standard error of an analysis assuming the same intraclass correlation coefficient ρ_{pooled} in both conditions (i.e., a pooled analysis) to an analysis assuming different intraclass correlation coefficient values (i.e., a separate analysis) is given by

$$\sqrt{\frac{2 + (\tilde{n}_1 + n_1 - 2)\rho_{pooled}}{2 + (n_1 - 1)\rho + (\tilde{n}_1 - 1)\tilde{\rho}}}; \qquad (8.2)$$

see Roberts and Roberts (2005). Here, n_1 and $\tilde{n}_1$ are the common cluster sizes in the experimental and control arms, respectively. The pooled estimate of the intraclass correlation coefficient is

$$\rho_{pooled} \cong \frac{(1 - 1/n_1)\rho + (1 - 1/\tilde{n}_1)\tilde{\rho}}{(1 - 1/n_1) + (1 - 1/\tilde{n}_1)}. \tag{8.3}$$

From this equation it follows that ρ_{pooled} is simply the average of ρ and $\tilde{\rho}$ when cluster sizes do not vary over treatment arms.

Combining Equations (8.2) and (8.3) the ratio of the standard error of a pooled analysis to a separate analysis can be calculated. For instance, if $n_1 = 10$, $\tilde{n}_1 = 40$, $\rho = 0.1$ and $\tilde{\rho} = 0.2$ then this ratio is 0.93, meaning that the standard error is underestimated by 7%. If $n_1 = 40$, $\tilde{n}_1 = 10$, $\rho = 0.1$ and $\tilde{\rho} = 0.2$ then this ratio is 1.09, and thus the standard error is overestimated by 9%. The ratio departs from 1 when the ratio $n_1/\tilde{n}_1$ departs from 1 and/or when the ratio $\rho/\tilde{\rho}$ departs from 1. In general, one can state that underestimation occurs when the intraclass correlation is lower for the arm with smaller cluster sizes, and that overestimation occurs when the intraclass correlation is larger for the arm with smaller cluster sizes.

Although an under- or overestimation of the standard error by a few per cent may seem trivial, it may have an effect on the p-value of the test on treatment effect and hence on the conclusion with respect to the effectiveness of the experimental condition. Implications for sample size issues are further discussed in Section 8.3.

8.2.2 Clustering in one treatment arm

Henceforward, the condition without clustering is referred to as the control condition and the condition with subjects in clusters as the experimental condition, although in some trials clustering may occur in the control condition and not in the experimental condition. Separate regression models in both conditions are formulated and then combined into one single equation. The model for subject i in the control condition is

$$y_i = \gamma_0 + r_i. \tag{8.4}$$

The error terms $r_i \sim N(0, \sigma_r^2)$ are assumed to be independently distributed as the subjects in the control condition are assumed not to influence each other's behavior.

The model for subjects in the experimental condition is more complicated because of the nesting of subjects within clusters. Such clusters may be therapy groups, risk reduction sessions or peer pressure groups, or a number of patients or clients treated by the same health professional or therapist.

The model for subject i in therapy group j is given by

$$y_{ij} = \gamma_0 + \gamma_1 + u_j + e_{ij}. \tag{8.5}$$

The parameter γ_0 is the mean outcome in the control condition and γ_1 is the treatment effect. This model has random effects at the subject and cluster level: $e_{ij} \sim N(0, \sigma_e^2)$ and $u_j \sim N(0, \sigma_u^2)$. The total variance of a subject's outcome in the experimental condition is $var(y_{ij}) = \sigma^2 = \sigma_u^2 + \sigma_e^2$ and, as usual, the proportion variation at the cluster level is defined by the intraclass correlation coefficient $\rho = \sigma_u^2/(\sigma_u^2 + \sigma_e^2)$. The total variance σ^2 does not necessarily have to be equal to the variance σ_r^2 in the control condition. Thus, the model allows for heteroscedasticity.

Models (8.4) and (8.5) can be combined into a single equation model

$$y_{ij} = \gamma_0 + \gamma_1 x_{ij} + u_j x_{ij} + e_{ij} x_{ij} + r_{ij}(1 - x_{ij}). \tag{8.6}$$

The treatment condition x_{ij} indicates whether a subject is assigned to the control ($x_{ij} = 0$) or experimental ($x_{ij} = 1$) condition. As in the previous subsection, this variable appears in both the fixed and random parts of the multilevel model. One can easily verify that models (8.4) and (8.5) can be obtained by setting x_{ij} equal to 0 and 1, respectively.

To specify model (8.6) in a computer program for multilevel analysis one should create a constant variable for the intercept, a dummy variable for treatment condition x_j and another dummy for its complement $1 - x_j$. The intercept variable should be assigned a fixed effect but no random effects. The dummy variable for x_j should be assigned a fixed effect and random effects at the individual and cluster level. The dummy variable for $1 - x_j$ should be assigned a random effect at the individual level but not at the cluster level, and it should not be assigned a fixed effect.

8.3 Sample size calculations for continuous outcomes

8.3.1 Clustering in both treatment arms

8.3.1.1 Factors that influence power

The null hypothesis $H_0 : \gamma_1 = 0$ is tested with the test statistic $t = \hat{\gamma}_1/s.e.(\hat{\gamma}_1)$ which, under the null hypothesis, follows a central t distribution. The degrees of freedom are not as easily calculated as for the designs in the previous chapters. Hoover (2002) provides a power formula on basis of the t test. He uses the Satterthwaite approximation (Satterthwaite, 1946) with a correction of the degrees of freedom derived by DiSantostefano and Muller (1995). Alternatively one may use an exact numerical method (Moser, Steves, & Watts, 1989), which is implemented in nQuery Advisor (Elashoff, 2007). For large sample size the t distribution may be approximated by the standard normal, but this procedure overestimates power when sample size is small.

Under the alternative hypothesis, the test statistic follows a noncentral t

distribution with non-centrality parameter

$$\lambda = \frac{\gamma_1}{s.e.(\hat{\gamma}_1)} = \frac{\gamma_1}{\sqrt{\frac{\sigma_e^2+n_1\sigma_u^2}{n_1 n_2} + \frac{\tilde{\sigma}_e^2+\tilde{n}_1\tilde{\sigma}_u^2}{\tilde{n}_1\tilde{n}_2}}}. \tag{8.7}$$

The cluster size, number of clusters per condition, total outcome variance and intraclass correlation coefficient in the experimental condition are denoted n_1, n_2, σ^2 and ρ. A tilde is used for their counterparts in the control condition. The standardized effect size is defined as $\delta = \gamma_1/\tilde{\sigma}$size, which expresses the treatment effect in terms of standard deviation of the outcome in the control condition. Given this definition, the non-centrality parameter can be rewritten as

$$\lambda = \frac{\delta}{\sqrt{\frac{\theta(1+(n_1-1)\rho)}{n_1 n_2} + \frac{1+(\tilde{n}_1-1)\tilde{\rho}}{\tilde{n}_1\tilde{n}_2}}} \tag{8.8}$$

where $\theta = \sigma^2/\tilde{\sigma}^2$ is the ratio of the total variance in the experimental condition relative to that in the control. This equation is an extension to that given by Roberts and Roberts (2005) since it accounts for differential total variance across the treatment arms. The argument of the square root at the right side of the equation consists of two terms that represent the variances of the mean outcomes in the experimental and control conditions, respectively. A covariance between these means is not included since it is assumed the scores in the experimental arm are independent of those in the control arm.

As in the previous chapters the non-centrality parameter, and hence power, increases with increasing effect size γ_1 and decreasing standard error $s.e.(\gamma_1)$. The standard error depends on the total outcome variance and intraclass correlation coefficients and on the number of clusters and cluster size in both treatment conditions. Increasing sample size results in increasing power level. As for cluster randomized trials, the number of clusters has a larger effect on power than the cluster size since cluster size appears in both the numerator and denominator while the number of clusters only appears in the denominator. Thus a small number of clusters in both arms cannot always be compensated by large cluster sizes to achieve a desired power level.

8.3.1.2 Sample size formulae for fixed cluster sizes

The clusters in individually randomized group treatment trials are often focus groups, peer pressure groups, therapists or health professionals. For practical reasons, the sizes of such clusters are often fixed in advance. Focus groups and peer pressure groups, for instance, are often small to promote dialogue among participants. Furthermore, focus group sessions may sometimes include exchange of sensitive information and participants might not be willing to participate in discussions when group size is too large. The number of patients or clients treated by the same health professional is often limited by the amount of time such a health professional is assigned for treatment sessions and preparation. For this reason, only formulae to calculate the required number of clusters in either arm are provided, given the assumption that all clusters within an arm have the same cluster size.

Given cluster sizes n_1 and $\tilde{n}_1$ and a fixed number of clusters in the experimental condition n_2, the required number of clusters in the control condition is

$$\begin{aligned}\tilde{n}_2 &= \frac{\tilde{\sigma}_e^2 + \tilde{n}_1\tilde{\sigma}_u^2}{\tilde{n}_1\left(\frac{\gamma_1^2}{(z_{1-\alpha}+z_{1-\beta})^2} - \frac{\sigma_e^2+n_1\sigma_u^2}{n_1 n_2}\right)} \\ &= \frac{1+(\tilde{n}_1-1)\tilde{\rho}}{\tilde{n}_1\left(\frac{\delta^2}{(z_{1-\alpha}+z_{1-\beta})^2} - \frac{\theta(1+(n_1-1)\rho)}{n_1 n_2}\right)}. \end{aligned} \tag{8.9}$$

As in the previous chapters, the calculation of sample size is based on the normal approximation, which works well when degrees of freedom are sufficiently large. A calculation based on the t distribution would require an iterative procedure since the degrees of freedom are a function of the number of clusters. Equation (8.9) shows that the number of control clusters increases with increasing power level and decreases with increasing effect size and type I error rate. Fewer clusters in the control condition are needed when the cluster size in the control condition increases and when the number of clusters and cluster size in the experimental condition increase. On the other hand, more clusters in the control condition are needed when the total variance and intraclass correlation in either condition increase. A similar expression can be given for the required number of clusters in the experimental condition by exchanging the symbols $\tilde{n}_1$, $\tilde{n}_2$, $\tilde{\rho}$ and $\tilde{\sigma}^2$ by their counterparts in the experimental condition and vice versa, and is not presented here.

The findings with respect to cluster size and number of clusters are illustrated in Figure 8.1. This figure shows the power to detect a small treatment effect in a test with a two-sided alternative hypothesis and $\alpha = 0.05$ and $\rho = \tilde{\rho} = 0.05$ and under the assumption of equal total outcome variance. Fewer clusters in the control condition are needed when the experimental arm consists of more clusters. Comparing the left and right sides of this figure shows that increasing the common cluster size from $n_1 = \tilde{n}_1 = 20$ to 40 has a small effect on power.

8.3.1.3 Including budgetary constraints

Equation (8.9) calculates the required number of clusters in the control condition when the number of clusters in the experimental condition is fixed. Such an approach may be useful, for instance, in situations where the experimental condition consists of group sessions and the number of groups is limited by the number of therapists who can supervise such groups. As explained above, Equation (8.9) can be rewritten to calculate the required number of clusters in the experimental condition, given a fixed number of control clusters. When neither the number of experimental and control clusters is fixed beforehand, a budgetary constraint may be used to calculate the optimal allocation ratio.

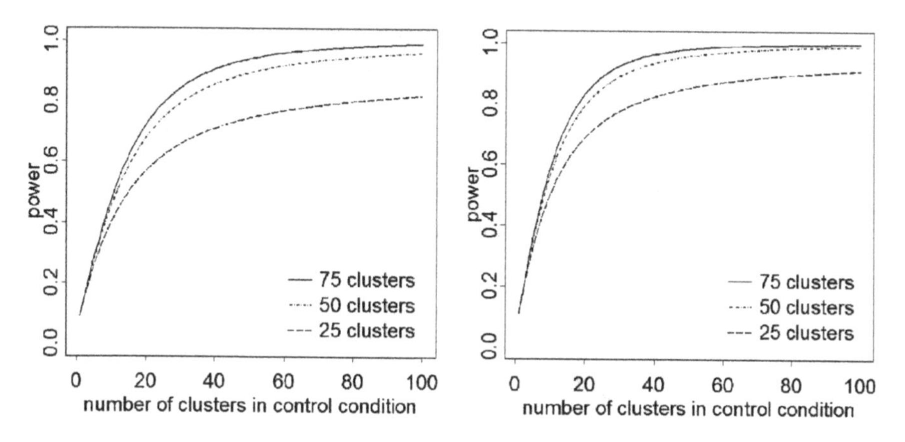

Figure 8.1: Power as a function of the number of clusters in both treatment conditions and for common cluster sizes 20 (left) and 40 (right).

This constraint is given by

$$C = c_E n_1 n_2 + c_C \tilde{n}_1 \tilde{n}_2, \tag{8.10}$$

where C represents the costs of the trial, and c_E and c_C are the costs per subject in the experimental and control conditions. Given this constraint, the costs to achieve a specific power level are minimized for the allocation ratio

$$r^* = \frac{n_1^* n_2^*}{\tilde{n}_1^* \tilde{n}_2^*} = \sqrt{\frac{c_C \sigma^2 (1 + (n_1 - 1)\rho)}{c_E \tilde{\sigma}^2 (1 + (\tilde{n}_1 - 1)\tilde{\rho})}}. \tag{8.11}$$

This equation reduces to Equation (3.21) if outcome measures are uncorrelated in both treatment arms (i.e., when $\rho = \tilde{\rho} = 0$). This ratio increases if the cost ratio c_C/c_E increases, the variance ratio $\sigma^2/\tilde{\sigma}^2$ increases and the ratio of the design effects $(1 + (n_1 - 1)\rho)/(1 + (\tilde{n}_1 - 1)\tilde{\rho})$ increases. So one would sample fewer control subjects in favor of sampling more experimental subjects when the costs c_C in the control arm are high relative to the costs c_E in the experimental arm, when the total outcome variance σ^2 in the experimental arm is high relative to the total outcome variance $\tilde{\sigma}^2$ in the control arm, and when the design effect $1 + (n_1 - 1)\rho$ in the experimental arm is large relative to the design effect $1 + (\tilde{n}_1 - 1)\tilde{\rho}$ in the control arm.

The optimal number of groups in the control condition to achieve a power level $1 - \beta$ is now calculated from

$$\tilde{n}_2^* = \frac{\sqrt{\frac{c_E}{c_C}}\sqrt{\sigma^2(1 + (n_1 - 1)\rho)}\sqrt{\tilde{\sigma}^2(1 + (\tilde{n}_1 - 1)\tilde{\rho})} + \tilde{\sigma}^2(1 + (\tilde{n}_1 - 1)\tilde{\rho})}{\tilde{n}_1 \left(\frac{\gamma_1}{z_{1-\alpha} + z_{1-\beta}}\right)^2}, \tag{8.12}$$

and the corresponding number of groups in the experimental condition follows from substitution of $\tilde{n}_2^*$ into Equation (8.11).

It is worthwhile to mention that Equation (8.10) can also be used to minimize the total number of subjects to achieve a specific power level. This is achieved by setting $c_E = c_C = 1$ so that the constraint reduces to $n_1 n_2 + \tilde{n}_1 \tilde{n}_2 = C$, where C is the total number of subjects. This approach is useful when a trial compares treatments for a rare disease or condition and the number of eligible subjects is expected to be small.

8.3.2 Clustering in one treatment arm

8.3.2.1 Factors that influence power

In case clustering occurs in just one arm, the non-centrality parameter is given by

$$\lambda = \frac{\gamma_1}{s.e.(\hat{\gamma}_1)} = \frac{\gamma_1}{\sqrt{\frac{\sigma_r^2}{n} + \frac{\sigma_e^2 + n_1 \sigma_u^2}{n_1 n_2}}} = \frac{\delta}{\sqrt{\frac{1}{n} + \frac{\theta(1+(n_1-1)\rho)}{n_1 n_2}}}, \tag{8.13}$$

where $\theta = \sigma^2/\sigma_r^2$ is the total outcome variance in the experimental condition relative to the control. The cluster size and number of clusters in the experimental condition are denoted n_1 and n_2, and n is the number of subjects in the control condition. As in the previous subsection, the standardized effect size expresses the treatment effect in terms of the standard deviation in the control condition: $\delta = \gamma_1/\sigma_r$.

A justification of the standard error is given in Moerbeek and Wong (2008). The argument of the square root of the denominator consists of two parts that represent the variances of the mean outcomes in the control and experimental conditions. The covariance of these two means is equal to zero since the scores of the outcome measures in the control condition are assumed to be independent of those in the experimental condition. We may expect to see dependency of outcome measurements among those in the same cluster in the experimental condition, which is represented by the intraclass correlation coefficient in the second argument of the square root.

The non-centrality parameter is a function of the effect size, the total variances in both treatment arms, the intraclass correlation coefficient and the sample sizes n_1, n_2 and n. Given fixed number of clusters and fixed cluster size in the experimental condition, the non-centrality parameter, and hence the power, increases with increasing number of control subjects. Given a fixed sample size in the control condition, the non-centrality parameter increases with increasing cluster size and number of clusters. As with cluster randomized trials, the effect of the number of clusters is larger than the effect of the cluster size.

8.3.2.2 Sample size formulae for fixed cluster sizes

Once again sample size formulae are derived under the restriction that the cluster size in the experimental condition is fixed. In some trials the number of clusters is fixed beforehand, for instance when the number of professionals who can supervise such clusters is limited. In that case, the optimal number of control subjects to achieve a power $1-\beta$ to detect a treatment effect γ_1 in a test with a one-sided alternative hypothesis $H_a: \gamma_1 < 0$ or $H_a: \gamma_1 > 0$ and type I error probability α is given by

$$\begin{aligned} n &= \frac{\sigma_r^2}{\left(\frac{\gamma_1^2}{(z_{1-\alpha}+z_{1-\beta})^2} - \frac{\sigma_e^2+n_1\sigma_u^2}{n_1 n_2}\right)} \\ &= \frac{1}{\left(\frac{\delta^2}{(z_{1-\alpha}+z_{1-\beta})^2} - \frac{\theta(1+(n_1-1)\rho)}{n_1 n_2}\right)}. \end{aligned} \tag{8.14}$$

For a two-sided alternative $H_a: \gamma_1 \neq 0$ the type I error rate α is replaced by $\alpha/2$. Fewer control subjects are required when the variance ratio θ and intraclass correlation coefficient ρ are small and/or when the number of clusters and cluster size in the experimental condition are large.

In other trials the number of control subjects is fixed beforehand and the number of clusters in the experimental condition is calculated as

$$n_2 = \frac{\sigma_e^2 + n_1\sigma_u^2}{n_1\left(\frac{\gamma_1^2}{(z_{1-\alpha}+z_{1-\beta})^2} - \frac{\sigma_r^2}{n}\right)} = \frac{\theta(1+(n_1-1)\rho)}{n_1\left(\frac{\delta^2}{(z_{1-\alpha}+z_{1-\beta})^2} - \frac{1}{n}\right)}. \tag{8.15}$$

Fewer clusters are required when the cluster size and number of control subjects are large, and/or when the variance ratio and intraclass correlation coefficient are small.

Figure 8.2 shows the power to detect a small treatment effect in a trial with a two-sided alternative hypothesis and $\alpha = 0.05$. The intraclass correlation coefficient in the experimental group was fixed at $\rho = 0.05$ and the outcome variance did not vary over treatment arms (i.e., $\theta = 1$). Fewer control subjects are required when the number of experimental clusters increases from 25 to 50. Further increasing the number of clusters to 75 has a smaller effect on power. Doubling the cluster size from 20 to 40 does not have such a large effect on power as doubling the number of clusters from 25 to 50.

It should be noted that both Equations (8.14) and (8.15) assume all clusters in the experimental condition have equal size. This assumption is not always met in practice and the efficiency loss due to varying cluster sizes rarely exceeds 10%. This can be compensated by recruiting 11% more clusters in the experimental arm and 11% more subjects in the control arm (Candel, 2009).

8.3.2.3 Including budgetary constraints

In some trials the sample sizes are not fixed beforehand and a budgetary constraint may be used to calculate the optimal allocation ratio and the required

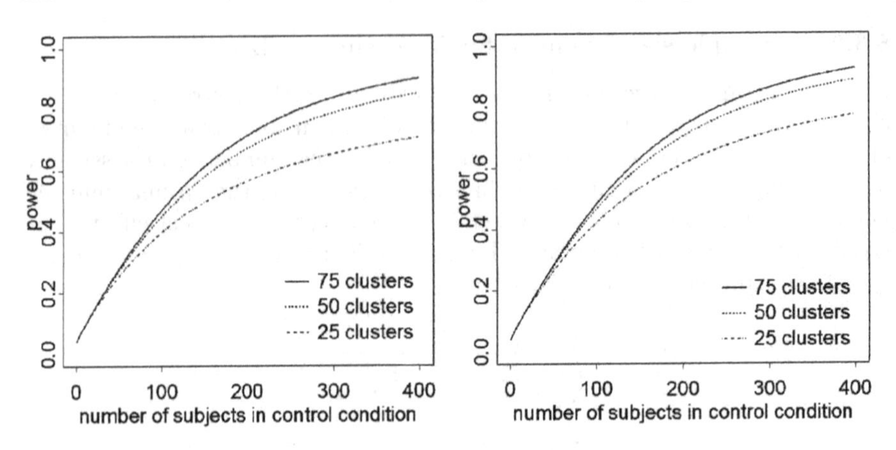

Figure 8.2: Power as a function of the number of control subjects, for different values of the number of clusters in the experimental condition and for cluster sizes 20 (left) and 40 (right).

sample sizes in both conditions to achieve a desired power level. For trials with clustering in one arm, the budgetary constraint is given by

$$C = c_E n_1 n_2 + c_C n, \tag{8.16}$$

where c_E and c_C are the costs per subject in the experimental and control condition, and C are the total costs. The costs in the experimental condition include costs to take measurements on subjects and also the costs of the health professional who supervises the groups in which the subjects in the experimental condition are nested (averaged over the number of subjects in the group). Given this relation, the optimal allocation ratio is given by

$$r^* = \frac{n_1^* n_2^*}{n} = \sqrt{\theta\,(1+(n_1-1)\,\rho)c_C/c_E}, \tag{8.17}$$

where once again the variance ratio $\theta = \sigma^2/\sigma_r^2$ is the ratio of the total variance in the experimental group to the variance in the control group. The number of subjects in the experimental condition, relative to the number of control subjects, increases with the variance ratio θ, the design effect $1 + (n_1 - 1)\rho$, and the cost ratio c_C/c_E. Thus, more subjects in the experimental condition are required when the variability of the outcomes in this condition increases and when the degree of dependency of outcomes within the same group increases. The latter relation is stronger for large group sizes. Fewer subjects in the experimental condition are needed when the costs for a subject in the experimental condition increase. Of course, increasing the sample size in the experimental condition implies that fewer control subjects are required to achieve the desired power level, and vice versa.

The required sample size in the control group to achieve a power $1 - \beta$ at smallest costs is given by

$$n^* = \left(\sqrt{\theta(1+(n_1-1)\rho)c_E/c_C}+1\right)\left(\frac{z_{1-\alpha}+z_{1-\beta}}{\delta}\right)^2. \tag{8.18}$$

Once n is available, it is imputed into Equation (8.17) to calculate the corresponding number of subjects in the experimental condition.

In case $c_E = c_C = 1$, the budgetary constraint reduces to $C = n_1 n_2 + n$, which implies the optimal allocation is such that the total number of subjects is minimized.

8.4 Sample size calculations for dichotomous outcomes

The sample sizes for individually randomized group treatment trials with dichotomous outcomes are based on the risk difference $p_C - p_E$. The event proportions in the control and experimental condition are denoted p_C and p_E, respectively. It is assumed sample sizes are large so that Equation (3.20) can be used to relate sample size to power.

8.4.1 Clustering in both treatment arms

For trials with clustering in both treatment conditions, the standard error of the risk difference is equal to

$$s.e.(\hat{p}_C - \hat{p}_E) = \sqrt{\frac{p_C(1-p_C)(1+(\tilde{n}_1-1)\tilde{\rho})}{\tilde{n}_1\tilde{n}_2} + \frac{p_E(1-p_E)(1+(n_1-1)\rho)}{n_1 n_2}}. \tag{8.19}$$

The terms n_1, n_2 and ρ denote the cluster size, number of clusters and intraclass correlation coefficient in the experimental condition and a tilde is used to indicate their counterparts in the control condition. As for continuous outcomes, the focus is on trials where the cluster sizes n_1 and $\tilde{n}_1$ are fixed. Given fixed cluster sizes and a fixed number of clusters in the experimental condition, the required number of clusters in the control condition is given by

$$\tilde{n}_2 = \frac{p_C(1-p_C)(1+(\tilde{n}_1-1)\tilde{\rho})}{\tilde{n}_1\left(\frac{(p_C-p_E)^2}{(z_{1-\alpha}+z_{1-\beta})^2} - \frac{p_E(1-p_E)(1+(n_1-1)\rho)}{n_1 n_2}\right)}. \tag{8.20}$$

The same equation can be used to calculate the required number of clusters in the experimental condition, given a fixed number of control clusters, by replacing n_1, n_2 and ρ by their counterparts in the control condition and vice versa.

The budget constraint (8.10) is used to calculate the optimal number of clusters in both arms when neither is fixed. The optimal allocation ratio is then given by

$$r^* = \frac{n_1^* n_2^*}{\tilde{n}_1^* \tilde{n}_2^*} = \sqrt{\frac{c_C p_E(1-p_E)(1+(n_1-1)\rho)}{c_E p_C(1-p_C)(1+(\tilde{n}_1-1)\tilde{\rho})}}. \tag{8.21}$$

The budget to achieve a power level $1-\beta$ is minimized when the number of clusters in the control condition is calculated from

$$\tilde{n}_2^* = \frac{\sqrt{\frac{c_E}{c_C}}\sqrt{\sigma^2(1+(n_1-1)\rho)}\sqrt{\tilde{\sigma}^2(1+(\tilde{n}_1-1)\tilde{\rho})} + \tilde{\sigma}^2(1+(\tilde{n}_1-1)\tilde{\rho})}{\tilde{n}_1\left(\frac{p_c-p_E}{z_{1-\alpha}+z_{1-\beta}}\right)^2}, \tag{8.22}$$

where $\sigma^2 = p_E(1-p_E)$ and $\tilde{\sigma}^2 = p_C(1-p_C)$ are the outcome variance in the experimental and control condition. The corresponding number of experimental clusters is then calculated from Equation (8.21).

8.4.2 Clustering in one treatment arm

For trials with clustering in just one treatment arm, the standard error of the risk difference is given by

$$s.e.(\hat{p}_C - \hat{p}_E) = \sqrt{\frac{p_C(1-p_C)}{n} + \frac{p_E(1-p_E)(1+(n_1-1)\rho)}{n_1 n_2}}. \tag{8.23}$$

The first term of the argument of the square root is the variance of the estimated proportion $\hat{p}_C$ in the control condition and the second term is the variance of the estimated proportion $\hat{p}_E$ in the experimental condition. The second term includes the design effect $1+(n_1-1)\rho$ and so the variance depends on the intraclass correlation coefficient and cluster size. This reduces to Equation (3.10) if the intraclass correlation coefficient is equal to zero.

Given a fixed cluster size and fixed number of clusters in the experimental condition, the required number of control subjects is equal to

$$n = \frac{p_C(1-p_C)}{\frac{p_E(1-p_E)(1+(n_1-1)\rho)}{n_1 n_2} + \frac{(p_C-p_E)^2}{(z_{1-\alpha}+z_{1-\beta})^2}}. \tag{8.24}$$

Similarly, the required number of clusters of size n_1 in the experimental condition can be calculated once the number of control subjects is fixed:

$$n_2 = \frac{p_E(1-p_E)(1+(n_1-1)\rho)}{n_1\left(\frac{p_C(1-p_C)}{n} + \frac{(p_C-p_E)^2}{(z_{1-\alpha}+z_{1-\beta})^2}\right)}. \tag{8.25}$$

If neither sample size is fixed, the budgetary constraint (8.16) is used to calculate the optimal allocation ratio

$$r^* = \frac{n_1^* n_2^*}{n^*} = \sqrt{\tau((n_1-1)\rho+1)c_C/c_E}, \tag{8.26}$$

where the quantity $\tau = p_E(1-p_E)/(p_C(1-p_C))$ is the ratio of the variances of the outcome in both arms. The number of subjects in the control condition to achieve a power $1-\beta$ at lowest cost is calculated from

$$n^* = p_C(1-p_C)(\sqrt{\tau((n_1-1)\rho+1)c_E/c_C}+1)\left(\frac{z_{1-\alpha}+z_{1-\beta}}{p_C-p_E}\right)^2, \quad (8.27)$$

and should be imputed in Equation (8.26) to calculate the required number of clusters in the experimental condition.

8.5 An example

Dolbeault et al. (2009) studied the effectiveness of a psycho-educational group therapy after early-stage breast cancer treatment in a randomized controlled trial. The control group consisted of a waiting list, so the design implies nesting in the group therapy arm but not in the control arm. Randomization to treatment conditions was done at the individual level. The group therapy program included eight weekly sessions that comprised thematic discussions, information and training in stress managements techniques. Groups consisted of 8 to 12 participants led by two therapists, either psychologists or psychiatrists. Participants within the same therapy group may be expected to mutually influence each others' behavior. In addition, variations in the skills and experience of the therapists may lead to therapist effects. It is therefore likely that outcomes of participants in the same therapy group are correlated. Unfortunately, the size of such correlation is not given in the paper by Dolbeault et al. (2009). As the control condition consisted of a waiting list, correlated outcomes in the control arm are not expected.

Measurements were taken at baseline, at the end of intervention and at one-month follow up. The group therapy and control condition consisted of 102 and 101 participants; at follow up measurements were taken on 81 and 87 participants, respectively. The paper by Dolbeault et al. (2009) studied 27 outcome variables, of which the interpersonal relationships outcome as measured at follow-up is used to illustrate the use of the sample size formulae in Section 8.3.2.

Interpersonal relationships is a continuous outcome variable (range 0 to 28) measured by means of the Profile of Mood States (POMS) rating scale (McNair, Lorr, & Droppleman, 1992). At follow-up, the mean in the group therapy arm was 18.80 with standard deviation 3.49, and the mean in the control arm was 17.41 with standard deviation 3.67. The standardized effect size, expressed in terms of standard deviation units in the control arm, is equal to $\delta = (18.80 - 17.41)/3.49 = 0.40$, which corresponds to a medium effect size. At the given sample sizes, the null hypothesis of no treatment effect would be rejected with $p = 0.0064$.

Suppose a researcher wishes to evaluate the effectiveness of the group therapy program in another hospital. To implement the formulae, parameter estimates from Dolbeault et al. (2009) are used: $\theta = \sigma^2/\sigma_r^2 = 3.67^2/3.49^2 = 1.11$ and $\delta = 0.40$. The common group size in the experimental arm is fixed to be equal to $n_1 = 10$ and an intraclass correlation coefficient $\rho = 0.05$ is expected. Sample size calculations are provided to achieve a power of 0.8 on a test with a one-sided alternative hypothesis and type I error rate $\alpha = 0.05$.

Recruiting subjects into the study is expected to be the main hurdle, and the researcher aims to design the study such that the number of subjects is minimized. Thus, Equation (8.16) is used with $c_E = c_C = 1$ as a constraint. On basis of Equation (8.17) the optimal allocation ratio is equal to $r^* = 1.27$, which implies 1.27 times as many subjects are required in the group therapy arm as in the control arm. The optimal number of control subjects is calculated from Equation (8.18) and is equal to $n^* = 87.66$. The total number of subjects in the therapy group condition is 1.27 times as large: $n_1 n_2 = 111.21$. The total required sample size is $n + n_1 n_2 = 198.88$. These values need to be rounded to integers such that the sample size in the therapy group condition is a multiple of the group size $n_1 = 10$. Rounding downward the total number of subjects in the therapy group condition to $n_1 n_2 = 110$ enables the researcher to include $n_2 = 11$ therapy groups. The corresponding sample size in the control arm is $n = 89$ and the total sample size is $n + n_1 n_2 = 199$. Rounding upward the total number of subjects in the therapy group condition to $n_1 n_2 = 120$ enables the researcher to include $n_2 = 12$ therapy groups. In that case, $n = 81$ control subjects are required and the resulting total sample size is $n + n_1 n_2 = 201$. The first scenario is favored since that results in slightly lower total sample size.

Figure 8.3 shows other values of the intraclass correlation coefficient result in different sample sizes. This figure shows that the sample size in the therapy group condition is stronger related to the intraclass correlation coefficient than the sample size in the control condition. If a slightly higher degree of correlation is assumed, say $\rho = 0.075$, then 217 subjects are needed. On the other hand, fewer subjects are needed if the intraclass correlation coefficient is lower. For instance, 182 subjects are needed if $\rho = 0.025$. Ignoring possible dependency of outcome within a therapy group altogether (i.e., assuming $\rho = 0$) results in an underpowered study with sample sizes $n = 84$, $n_1 n_2 = 80$ and a total sample size $n + n_1 n_2 = 164$. It is therefore important to include within-group variability in sample size calculations when such variability exists.

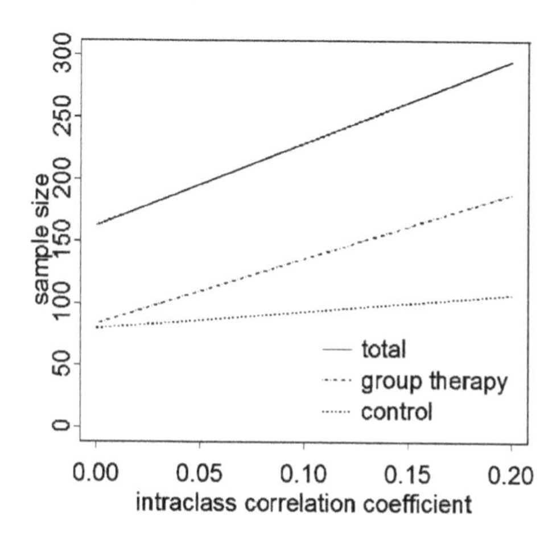

Figure 8.3: Required sample sizes for the outcome variable interpersonal relationships as a function of the intraclass correlation coefficient.

9

Longitudinal intervention studies

9.1 Introduction

The previous chapters focused on trials where the outcome variable was measured at one point in time post treatment. In this chapter an extension is made to trials where multiple measurements across time are taken, for instance, at pretest, at posttest and at one or more follow-up occasions. Such a trial is called a longitudinal intervention study and its data have a multilevel data structure with repeated measurements nested within subjects. As measurements within a given subject are likely to be correlated, the multilevel model is the appropriate model for data analysis. The next section gives a short introduction to the multilevel model for longitudinal data; for a more extensive introduction the reader is referred to Hedeker and Gibbons (2006) and Singer and Willett (2003).

The multilevel model can be shown to be the same model as the latent growth curve model within the framework of structural equation modeling (Duncan, Duncan, & Strycker, 2006). Both approaches have advantages over more traditional approaches for longitudinal data analysis, such as repeated measures (multivariate) analysis of variance: they are not restricted to constant (co-)variances across the measurement occasions; they do not require all subjects to be measured at the same points in time; intermittently missed observations or drop-out can easily be taken into account without having to remove all cases with incomplete data; and they are more powerful to detect group differences at least in linear growth (Fan, 2003; Fan & Fan, 2005). For this reason the multilevel model is used to relate the outcome to relevant explanatory variables and to derive the optimal design.

The multilevel model assumes that baseline measurements and growth curves vary across subjects. Instead of estimating each subject's growth curve, an average growth curve and the between-subject variability around the average growth curve are estimated. Predictors at the subject and repeated measures levels can be used to explain part of the between-subject variability in growth trajectories.

In experimental settings it is common practice to study whether the treatment condition has an effect on growth rates. For instance, one may wish to evaluate whether a new program to lose body weight results in the same rate of weight decline as an old program. To achieve maximal power on the test on treatment effect, the optimal design should be chosen among all possible designs.

Designs for longitudinal intervention studies prescribe the number of subjects, the number of measurement occasions and the duration of the study. So, decisions should be made to sample and treat few subjects and measure them often or to sample many subjects and measure them only a few times. Of course, statistical power to detect an effect of treatment increases with increasing sample size, frequency of observation and study duration but in practice these quantities are often limited by practical and financial considerations. For this reason we demonstrate how various designs can be compared on the basis of their costs for sampling and treating subjects and for taking repeat measurements.

A common feature of longitudinal intervention studies is that they are hampered by missing data caused by intermittently missed observations and drop-out. There are various reasons why not all subjects respond at all time points. For instance, respondents do not show up at all measurement occasions, they refuse to respond to some items in a questionnaire or the person who measures the subjects forgets to measure some or all outcome variables. Subjects who are allocated to the intervention condition may refuse to participate at some point because they find the effects of the promising new treatment disappointing or they experience adverse side effects, and those randomized to the control condition may drop out because they are randomized to the least interesting condition.

In addition, subjects may be lost to follow-up, for instance because they move to another location and change therapists, clinics or general practices. It is important that model parameters, such as the difference in growth trajectories across time across treatment groups, are equal for subjects who drop out during the course of the study and those who do not. This type of missing data is often referred to as missing at random (Schafer & Graham, 2002). If the missingness is non-ignorable, incorrect conclusions on the effect of treatment may be drawn. The designs in this chapter are derived under the assumption of missing at random. Missing data results in a lower power level but the power is larger than in an analysis that only uses data of subjects who have responded on all measurement occasions (i.e., a complete case analysis).

This chapter restricts discussion to multilevel models for longitudinal data with two levels of nesting: repeated measurements within subjects. The extension to models that also include a cluster level is studied in the next chapter. In Section 9.2 the multilevel model for longitudinal data with continuous outcomes is given. Section 9.3 presents the relationship between statistical power and number of subjects, number of measurement occasions and study duration for trials without drop-out and compares various designs on basis of their costs. The effect of various drop-out patterns is studied in Section 9.4. The designs are derived under the assumption that measurements are taken at equidistant points in time. Such designs are very useful in practice since measurements taken at regular points in time decrease the risk of administrative

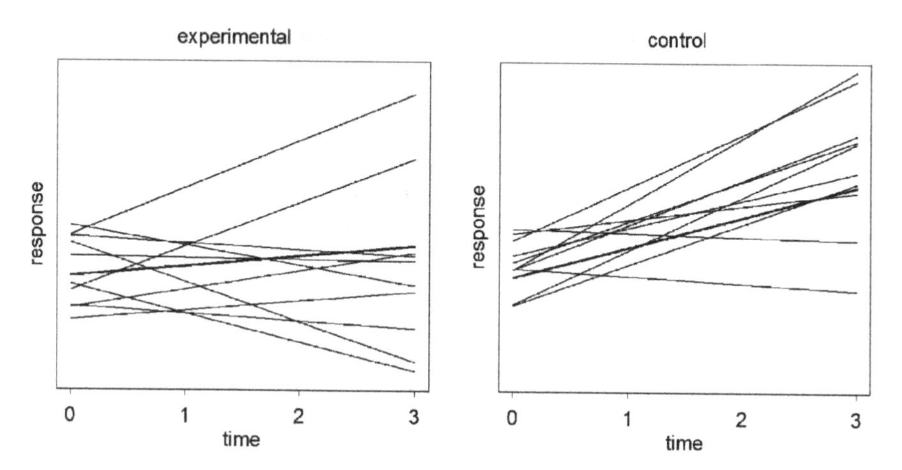

Figure 9.1: Linear relation between time and response. The bold lines represent the mean relations.

and logistic mistakes. In addition, designs with equidistant time points have been shown to be nearly as efficient as designs with an optimal allocation of time points (Tan & Berger, 1999).

9.2 Multilevel model

The basic idea of multilevel modeling for longitudinal data is that all subjects are assumed to have their own baseline score and growth patterns. This is shown on basis of a hypothetical example in Figure 9.1 that assumes a linear relation between time and response. The left and right panels correspond to the experimental and control conditions, respectively. For both conditions growth curves are shown for ten subjects and the mean relationships are plotted by means of bold lines. The mean baseline score hardly differs across treatment conditions, but a higher increase in response across time is observed in the control condition. It is observed that most subjects increase their responses across time but a negative relation between time and response is found for a few.

Instead of estimating each subject's growth curve, the mean growth curve is estimated along with the between-subject variability with respect to growth. To do so, a multilevel model is formulated that relates the response y_{ti} of subject $i = 1, ..., n_2$ at time point $t = 1, ..., n_1$ to the time point $time_t$ at which the measurement was taken:

$$y_{ti} = \pi_{0i} + \pi_{1i} time_t + e_{ti}. \tag{9.1}$$

As all subjects are measured at the same time points, the subject identifier i does not appear in the subscript of $time_t$. The response y_{ti} is assumed continuous and a linear multilevel model is used. The random error term e_{ti} follows a normal distribution with zero mean and variance σ_e^2. The baseline score π_{0i} and rate of change π_{1i} are assumed to vary across subjects and covariates at the subject level can be used to explain part of the between-subject variability, for instance treatment condition x_i:

$$\begin{aligned} \pi_{0i} &= \beta_{00} + \beta_{01} x_i + u_{0i} \\ \pi_{1i} &= \beta_{10} + \beta_{11} x_i + u_{1i}. \end{aligned} \quad (9.2)$$

Treatment condition is constant across time and hence the model does not allow for crossover. The random error terms u_{0i} and u_{1i} represent unexplained variability with respect to the intercept and slope. They are assumed normally distributed with zero mean, variances σ_{u0}^2 and σ_{u1}^2 and covariance σ_{u01}. In addition, they are assumed independent from the error term e_{ti} at the repeated measures level. A positive covariance implies that on average those with a higher baseline score increase their responses at a higher rate than those with lower baseline scores, and a negative covariance implies the opposite. A further reduction of unexplained variability can be achieved by adding time-varying predictors to the repeated measures level and time-invariant predictors to the subject level.

Substitution of Equation (9.2) in Equation (9.1) yields the single equation model

$$y_{ti} = \beta_{00} + \beta_{01} x_i + \beta_{10} time_t + \beta_{11} time_t x_i + u_{0i} + u_{1i} time_t + e_{ti}. \quad (9.3)$$

This model includes a cross-level interaction between time and treatment condition, which implies the mean linear change varies across treatments. To test the null hypothesis of equal mean growth across treatments, a statistical test should be performed for the regression weight β_{11}.

For linear models, the time variable is often rescaled such that the first measurement is taken at time point 0. In such a way, the intercept π_{0i} is the score of subject i at baseline. The vector *time* is the collection of time coefficients that describe the points in time at which measurements are taken. For instance, if the duration of the study is four years and measurements are taken once a year then $time = (0, 1, 2, 3, 4)'$ and for a study with the same duration and measurements once each two years $time = (0, 2, 4)'$.

In many applications the relation between time and response is curvilinear. An example is the relation between age and body length: the increase in body length decreases to zero at the end of adolescence. Polynomial regression is often used to model curvilinear relations between time and response. A first order polynomial model describes a linear relation between time and response, and by adding squares, cubes and higher order powers of the time coefficients as explanatory variables to model (9.1), quadratic, cubic and higher order polynomial relationships between time and response are modeled.

The problem with this approach is that the time coefficients and their powers are highly correlated, which may lead to estimation problems as a result of multicollinearity. For instance, the time vector $time = (0, 1, 2, 3, 4)'$ and its quadratic counterpart $time^2 = (0, 1, 4, 9, 16)'$ have a correlation of 0.96. The problem of multicollinearity can be solved by replacing the time coefficients by orthogonal polynomial contrast coefficients. For instance, the time vector and its quadratic counterpart are replaced by the vectors $c_1 = (-2, -1, 0, 1, 2)'$ and $c_2 = (1, -0.5, -1, -0.5, 1)'$. It can be easily verified than these two vectors are orthogonal, that is, $\sum_{t=1}^{n_1} c_{1t}c_{2t} = 0$, where n_1 is the number of measurement occasions including the baseline.

As one of the aims of this chapter is to study the effect of study duration on statistical power, orthogonal rather than orthonormal polynomials are used. With orthonormal polynomials, $\sum_{t=1}^{n_1} c_{pt}^2 = 1$, irrespective of study duration. For instance, the orthogonal polynomial coefficients c_1 as given above change to $c_1 = (-4, -2, 0, 2, 4)'$ if study duration doubles and measurements are taken at $time = (0, 2, 4, 6, 8)'$. The orthonormal coefficients are $c_1 = (-2, -1, 0, 1, 2)'/\sqrt{10}$ and do not change when study duration changes.

Raudenbush and Liu (2001) describe an algorithm to generate the orthogonal polynomial contrast coefficients and tables are listed in many textbooks. They can also be generated by the function `poly` in R or the procedure IML is SAS, for instance. A listing of linear, quadratic and cubic coefficients for trials with up to $n_1 = 12$ measurements per subject is given in Table 9.1. It is assumed any two adjacent measurements are taken one unit in time apart, that is, $time = (0, 1, ..., n_1 - 1)'$. It should be noted that for any number of measurements n_1 the coefficients c_{1p} are centered across zero. This implies that, in a linear model, the intercept π_{0i} represents the response of subject i at the midpoint of time rather than at baseline. The last column of Table 9.1 lists the summations $\sum_{t=1}^{n_1} c_{pt}^2$ which, as will be shown in the next section, are related to the standard error of the estimator of the between treatment difference in growth.

To describe a P-th order polynomial relation between time and response using orthogonal polynomial contrast coefficients, model (9.1) is replaced by the model

$$y_{ti} = \sum_{p=0}^{P} \pi_{pi} c_{pt} + e_{ti}, \tag{9.4}$$

where the individual change coefficients π_{pi} vary across subjects. As previously, treatment condition x_i can be used to model the between-subject variability:

$$\pi_{pi} = \beta_{p0} + \beta_{p1} x_i + u_{pi}. \tag{9.5}$$

The subject level error term u_{pi} is assumed to follow a normal distribution with zero mean and variance σ_{up}^2 and the covariance between error terms u_{pi} and $u_{p'i}$ is $\sigma_{up'p}$. The error terms u_{pi} are assumed independent from the error term e_{ti}. If treatment condition is coded $x_i = -0.5$ for the control and $x_i = 0.5$ for the experimental subjects, then β_{p1} is the mean difference between both

Table 9.1: Orthogonal polynomial contrast coefficients

n_1	c_p	$\sum_{t=1}^{n_1} c_{pt}^2$
3	$c_0' = (1,1,1)$	3
	$c_1' = (-1,0,1)$	2
4	$c_0' = (1,1,1,1)$	4
	$c_1' = (-3,-1,1,3)/2$	5
	$c_2' = (1,-1,-1,1)/2$	1
	$c_3' = (-1,3,-3,1)/20$	0.05
5	$c_0' = (1,1,1,1,1)$	5
	$c_1' = (-2,-1,0,1,2)$	10
	$c_2' = (2,-1,-2,-1,2)/2$	3.5
	$c_3' = (-1,2,0,-2,1)/5$	0.4
6	$c_0' = (1,1,1,1,1,1)$	6
	$c_1' = (-5,-3,-1,1,3,5)/2$	17.5
	$c_2' = (5,-1,-4,-4,-1,5)/3$	28/3
	$c_3' = (-5,7,4,-4,-7,5)/10$	1.8
7	$c_0' = (1,1,1,1,1,1,1)$	7
	$c_1' = (-3,-2,-1,0,1,2,3)$	28
	$c_2' = (5,0,-3,-4,-3,0,5)/2$	21
	$c_3' = (-1,1,1,0,-1,-1,1)$	6
8	$c_0' = (1,1,1,1,1,1,1,1)$	8
	$c_1' = (-7,-5,-3,-1,1,3,5,7)/2$	42
	$c_2' = (7,1,-3,-5,-5,-3,1,7)/2$	42
	$c_3' = (-7,5,7,3,-3,-7,-5,7)/4$	16.5
9	$c_0' = (1,1,1,1,1,1,1,1,1)$	9
	$c_1' = (-4,-3,-2,-1,0,1,2,3,4)$	60
	$c_2' = (28,7,-8,-17,-20,-17,-8,7,28)/6$	77
	$c_3' = (-28,14,26,18,0,-18,-26,-14,28)/10$	36.6
10	$c_0' = (1,1,1,1,1,1,1,1,1,1)$	10
	$c_1' = (-9,-7,-5,-3,-1,1,3,5,7,9)/2$	82.5
	$c_2' = (6,2,-1,-3,-4,-4,-3,-1,2,6)$	132
	$c_3' = (-42,14,35,31,12,-12,-31,-35,-14,42)/10$	85.86
11	$c_0' = (1,1,1,1,1,1,1,1,1,1,1)$	10
	$c_1' = (-5,-4,-3,-2,-1,0,1,2,3,4,5)$	110
	$c_2' = (15,6,-1,-6,-9,-10,-9,-6,-1,6,15)/2$	214.5
	$c_3' = (-30,7,22,23,14,0,-14,-23,-22,-7,30)/5$	171.6
12	$c_0' = (1,1,1,1,1,1,1,1,1,1,1,1)$	12
	$c_1' = (-11,-9,-7,-5,-3,-1,1,3,5,7,9,11)/2$	143
	$c_2' = (55,25,1,-17,-29,-35,-35,-29,-17,1,25,55)/6$	333 2/3
	$c_3' = (-33,3,21,25,19,7,-7,-7,-25,-21,-3,33)/4$	321.75

treatment groups with respect to polynomial effect of order p. The single equation model is obtained from substitution of (9.5) into (9.4):

$$y_{ti} = \sum_{p=0}^{P} (\beta_{p0} + \beta_{p1} x_i + u_{pi}) c_{pt} + e_{ti}, \qquad (9.6)$$

The interpretation of the individual change parameters π_{pi} depends on the order of the polynomial. For a linear growth curve (P=1), π_{1i} is the rate of change. For quadratic growth (P=2), π_{1i} is the average rate of change and π_{2i} represents acceleration. For cubic growth (P=3), π_{2i} is the average rate of acceleration and π_{3i} is the rate at which acceleration changes. In other words, adding higher order terms to a polynomial regression model changes the interpretation of lower order change parameters.

9.3 Sample size calculations for continuous outcomes

Power calculations in this section are based on the papers by Galbraith and Marschner (2002), Moerbeek (2008) and Raudenbush and Liu (2001). The first paper studies the effect of the number of subjects and number of measurement occasions on the power to detect differences between treatments for a fixed study duration for linear growth models. The latter papers extend to higher order polynomial growth and also study the effect of study duration on power. The costs of sampling and treating subjects and taking measurements are taken into account by the first two papers but not by the last.

A simple equation for the relation between power and design factors can be derived and is studied in this section. Design factors are the number of subjects, the number of measurement occasions, and study duration. It is shown how the number of subjects is calculated for a trial with a given duration and number of measurement occasions. In addition, costs are taken into account to compare various designs with the same level of power in terms of cost efficiency.

9.3.1 Factors that influence power

A main question in trials with longitudinal outcomes is whether a difference exists with respect to polynomial growth. The test statistic $t = \hat{\beta}_{p1}/s.e.(\hat{\beta}_{p1})$ is used to test the null hypothesis $H_0 : \beta_{p1} = 0$ against the alternative $H_1 : \beta_{p1} \neq 0$. Under the null hypothesis this test statistic follows a central t distribution with $n_2 - 2$ degrees of freedom and when the alternative is true it follows a non-central t distribution with non-centrality parameter

$$\lambda = \frac{\beta_{p1}}{\sqrt{\frac{4(\sigma_e^2 + \sigma_{up}^2 \sum_{t=1}^{n_1} c_{pt}^2)}{n_2 \sum_{t=1}^{n_1} c_{pt}^2}}}. \tag{9.7}$$

The unstandardized effect size appears in the numerator of this ratio; the denominator is the standard error

$$s.e.(\hat{\beta}_{p1}) = \sqrt{\frac{4(\sigma_e^2 + \sigma_{up}^2 \sum_{t=1}^{n_1} c_{pt}^2)}{n_2 \sum_{t=1}^{n_1} c_{pt}^2}}. \tag{9.8}$$

The power for the statistical test on β_{p1} is calculated as

$$\begin{array}{rl} P(t_{n_2-2,\lambda} > t_{n_2-2,1-\alpha}) & \text{when } H_a : \beta_{p1} > 0 \\ P(t_{n_2-2,\lambda} < t_{n_2-2,\alpha}) & \text{when } H_a : \beta_{p1} < 0 \\ P(t_{n_2-2,\lambda} > t_{n_2-2,1-\alpha/2}) + P(t_{n_2-2,\lambda} < t_{n_2-2,\alpha/2}) & \text{when } H_a : \beta_{p1} \neq 0 \end{array}. \tag{9.9}$$

The random variable $t_{n_2-2,\lambda}$ follows a non-central t distribution with $n_2 - 2$ degrees of freedom and non-centrality parameter λ and $t_{n_2-2,\alpha}$ and $t_{n_2-2,1-\alpha}$ are the 100αth and $100(1-\alpha)$th percentiles from the central t distribution with $n_2 - 2$ degrees of freedom.

In (9.7) the non-centrality parameter λ is expressed as a function of the unstandardized effect size β_{p1} but it may also be expressed in terms of a standardized, scale free, effect size. Raudenbush and Liu (2001) define the standardized effect size for polynomial trend p as $\delta_p = \beta_{p1}/\sigma_{up}$, which is the difference in polynomial trend p divided by its population standard deviation.

Consider as an example a trial that aims to evaluate the effects of an intervention to diminish anxiety among psychiatric patients. If the relation between time and anxiety is assumed to be linear, then an effect $\delta_1 = 0.5$ implies the mean difference between the control and intervention increases with 0.5 standard deviations σ_{u1} for each time unit. Imputation of the standardized effect size in the non-centrality parameter results in

$$\lambda = \frac{\delta_p}{\sqrt{\frac{4(\sigma_e^2/\sigma_{up}^2 + \sum_{t=1}^{n_1} c_{pt}^2)}{n_2 \sum_{t=1}^{n_1} c_{pt}^2}}}. \tag{9.10}$$

From Equations (9.7) through (9.10) it is observed that the power depends on the type I error rate α, the size of the treatment effect β_{p1} (or δ_p), the variance components σ_e^2 and σ_{up}^2 (or their ratio σ_e^2/σ_{up}^2), the number of subjects n_2, and the sum $\sum_{t=1}^{n_1} c_{pt}^2$. The latter is a function of the number of measurement occasions n_1 and the study duration d since the combination of the two determines the values of the orthogonal polynomial contrast coefficients c_{pt}.

Decreasing the term in the denominator of Equation (9.7) or (9.10) by using a more efficient design results in an increase of the power level. The design is determined by the number of subjects, the number of measurement occasions and the duration of the study. The effects of these three design factors and the variance ratio σ_e^2/σ_{up}^2 on power are now discussed. The related graphs show power levels to detect a medium effect size in a linear growth model for a trial with a two-sided alternative hypothesis and $\alpha = 0.05$.

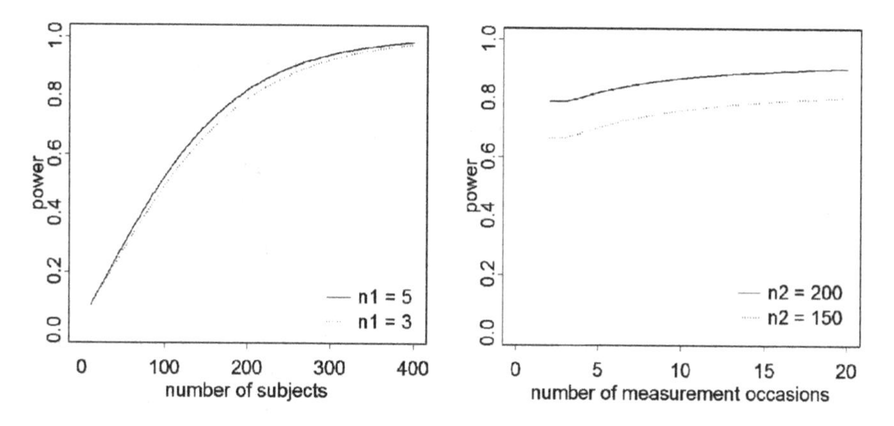

Figure 9.2: Power as a function of number of subjects and number of measurement occasions.

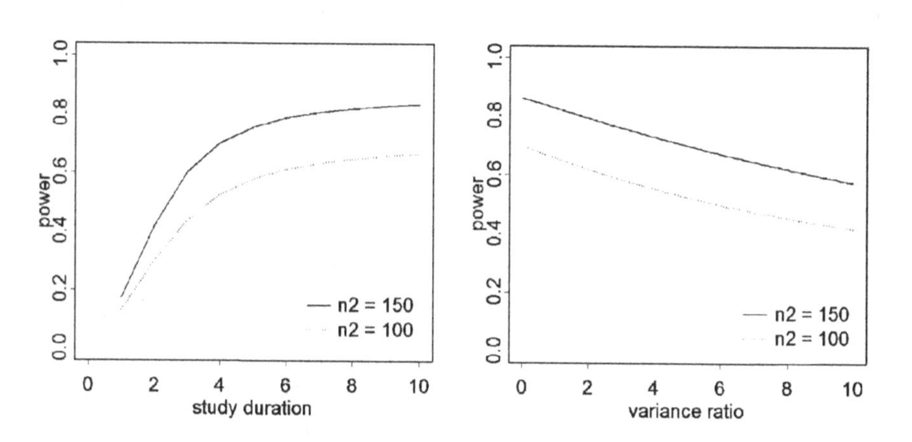

Figure 9.3: Power as a function of study duration and variance ratio.

Number of subjects. The number of subjects n_2 appears in the denominator of the standard error $s.e.(\hat{\beta}_{p1})$ but not in the numerator. Increasing n_2 to infinity implies the standard error approaches zero and hence the power approaches the value 1. As a result, for any number of measurements per subject and for any study duration one can achieve a desired power level by using a sufficiently large number of subjects. This is also depicted in the left panel of Figure 9.2 which shows the power as a function of the number of subjects (and for σ_e^2/σ_{up}^2=5, $d = 4$, and $n_1 = 3$ or $n_1 = 5$). The number of measurements per subject n_1 has a minor effect on power and for both values n_1 a power level of 0.8 is achieved with a sample size of about $n_2 = 200$. Increasing to values near $n_2 = 400$ results in a power near 1.0.

Number of measurement occasions. Increasing the number of measurement occasions n_1 while keeping study duration d constant implies a higher frequency of observation and hence a higher sum $\sum_{t=1}^{n_1} c_{pt}^2$. This sum not only appears in the denominator but also in the numerator of $s.e.(\hat{\beta}_{p1})$ and this standard error approaches the value $\sqrt{4\sigma_{up}^2/n_2}$ if the sum increases to infinity. Consequently, the desired power level cannot always be achieved by increasing the number of measurement occasions while keeping study duration constant. This finding is also visualized in the right panel of Figure 9.2 which shows power levels as a function of the number of measurement occasions for σ_e^2/σ_{up}^2=5, $d = 4$, and $n_2 = 150$ or $n_2 = 200$. Increasing n_1 from 2 to 3 does not have an impact on power. This can be seen by comparing the orthogonal polynomial contrast coefficients, which are $c_1 = (-2, 2)$ for two time points and $c_1 = (-2, 0, 2)$ for three time points. Both result in the same value of $\sum_{t=1}^{n_1} c_{pt}^2$ and hence the same power level. Further increasing n_1 has a minor effect on power and for $n_2 = 150$ a sufficient power level cannot be achieved by increasing n_1 alone.

Study duration. Increasing study duration d without changing the number of measurement occasions n_1 implies a lower frequency of observation. Multiplication of study duration d by a factor a while keeping the number of measurement occasions n_1 constant results in the multiplication of the sum $\sum_{t=1}^{n_1} c_{pt}^2$ by a factor a^{2p} (Raudenbush & Liu, 2001). So increasing d implies a higher $\sum_{t=1}^{n_1} c_{pt}^2$ and decreasing d implies this sum decreases. The exponent in a^{2p} depends on p, hence the larger the degree p of the polynomial effect the interest focuses on, the larger the effect of changing study duration. For the same reasoning that was given for the number of measurement occasions, a desired power level cannot always be achieved by increasing study duration to a sufficiently large value. See for an example the left panel of Figure 9.3 that shows increasing study duration does not result in a power of 0.8 for $n_1 = 100$ in a trial with σ_e^2/σ_{up}^2=5, and $n_1 = 5$.

Variance ratio. It is observed $s.e.(\hat{\beta}_{p1})$, and hence power, depends on the variance ratio σ_e^2/σ_{up}^2. The higher the ratio of the within- to the between-subject level variance of polynomial trend p, the higher the $s.e.(\hat{\beta}_{p1})$ and hence the lower the power, as is shown in an example in the right panel of Figure 9.3 ($n_1 = 5$, n_2= 100 or 150, and d=4). If one is able to eliminate all within-subject variability then the variance component σ_e^2 and also the variance ratio σ_e^2/σ_{up}^2 become zero. In that case the non-centrality parameter λ depends on δ_p and n_2 but not on the design factors n_1 and d.

9.3.2 Sample size formula for fixed number of measurements

The design of a longitudinal trial is determined by the number of subjects, the number of measurement occasions and study duration. If two of these design factors are fixed then the third can be calculated such that the desired power level is achieved. For instance, the number of measurement occasions may be limited if the amount of burden on the respondents should be minimized and the study duration may be limited if the results of the trial should become available within a certain time limit. In that case a sufficient power can be achieved by using a sample size at the subject level that is sufficiently large.

It is not straightforward to derive an equation for the number of subjects on basis of (9.9) since the degrees of freedom of the t distribution and non-centrality parameter are a function of this design factor. For large degrees of freedom the t distribution may be approximated by the standard normal. In that case, an explicit equation for the required number of subjects can be derived:

$$\begin{aligned} n_2 &= 4\frac{\sigma_e^2 + \sigma_{up}^2 \sum_{t=1}^{n_1} c_{pt}^2}{\sum_{t=1}^{n_1} c_{pt}^2} \left(\frac{z_{1-\alpha} + z_{1-\beta}}{\beta_{p1}} \right)^2 \\ &= 4\frac{\sigma_e^2/\sigma_{up}^2 + \sum_{t=1}^{n_1} c_{pt}^2}{\sum_{t=1}^{n_1} c_{pt}^2} \left(\frac{z_{1-\alpha} + z_{1-\beta}}{\delta_1} \right)^2. \end{aligned} \tag{9.11}$$

This will produce a non-integer value that has to be rounded upward to the nearest even value since the designs are based on the assumption that both treatment groups are of equal size. This equation hold for a one-sided alternative hypothesis. A larger sample size is needed for a two-sided alternative since α is replaced by $\alpha/2$.

As $s.e.(\hat{\beta}_{p1})$ depends on the study duration d and number of measurement occasions n_1 through $\sum_{t=1}^{n_1} c_{pt}^2$, a simple closed form relation between power and these two design factors cannot be given. To calculate the required study duration one can calculate the power for a range of feasible values d and select the smallest value d that results in the desired power level. A similar approach can be used for calculating the number of measurement occasions n_1.

9.3.3 Including budgetary constraints

Various combinations of the design factors result in the same level of power. For instance, the right panel of Figure 9.2 shows that a power of 0.8 can be achieved with $n_1 = 4$ measurements per subject and $n_2 = 200$ but also with $n_1 = 19$ and $n_2 = 150$. It is then good practice to select the design with lowest costs. Here the costs C are specified as a function of the number of subjects n_1 and the number of measurement occasions n_2: $C = c_1 n_1 n_2 + c_2 n_2$, where c_1 and c_2 are the costs of taking a measurement and the costs of sampling and treating a subject. The latter are averaged across both treatment conditions since both treatment groups are of equal size. Alternatively the costs may be expressed in units c_1: $n_1 n_2 + r n_2$, where $r = c_2/c_1$ is the ratio of the costs at the subject level to the costs at the measurements level. If $r < 1$ it is more expensive to take a measurement than to include a subject and vice versa if $r > 1$.

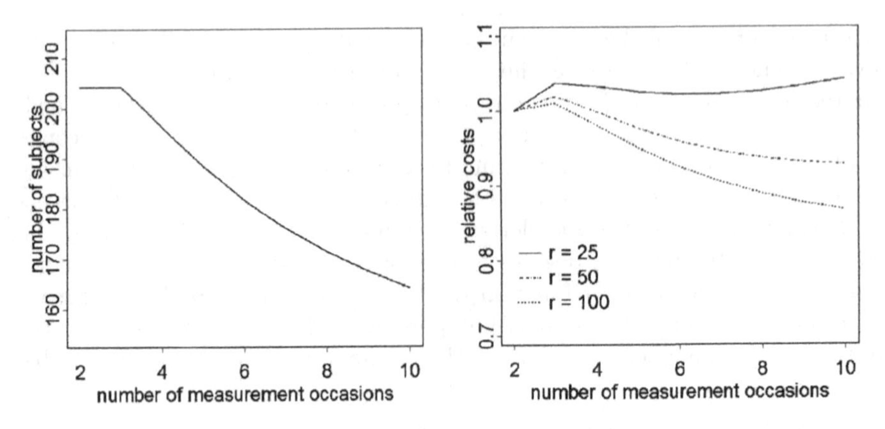

Figure 9.4: Number of subjects and relative costs as a function of number of measurement occasions in a trial without drop-out.

The left panel of Figure 9.4 shows the number of subjects as a function of the number of measurement occasions required to detect a medium effect size in a linear growth curve. A two-sided alternative hypothesis is used and a type I error rate $\alpha = 0.05$. The duration is fixed at $d = 4$ and a variance ratio $\sigma_e^2/\sigma_{up}^2 = 5$ is used. The number of measurement occasions is limited to $n_1 = 10$. The number of subjects does not change if n_1 increases from 2 to 3 but a smaller number of subjects is needed if the number of measurement occasions further increases. The right panel of Figure 9.4 shows the costs relative to the costs with $n_1 = 2$ as a function of n_1. It is observed the costs increase if n_1 increases from 2 to 3. For costs ratios $r = 25$ minimal costs are achieved at $n_1 = 2$ and for the other two costs ratios minimal costs are achieved at $n_1 = 10$. This is a logical finding since one would prefer a design with few time points and many subjects if it is inexpensive to sample a subject relative to taking a measurement. Of course the results in Figure 9.4 depend on the variance ratio and study duration and similar graphs should be made for other values to select the least expensive design.

9.4 Sample size calculations for dichotomous outcomes

Dichotomous outcomes are common in longitudinal research. In a smoking prevention intervention, for instance, researchers may want to measure the participants' smoking status (smoker versus non-smoker) at each measurement occasion. Another example comes from educational sciences, where participants may pass or fail a test at multiple occasions over time. The focus of this section is on the odds ratio. As will be explained below a simple mathematical expression of the power as a function of the number of subjects, the number of measurements and study duration cannot be given. For that reason, a simulation study is proposed to determine the power of a given design.

9.4.1 Odds ratio

At each measurement occasion the binary response y_{ti} is measured and coded 0 or 1. The probability p_{ti} of a score $y_{ti} = 1$ is a function of treatment condition x_i and time point $time_t$:

$$p_{ti} = \frac{1}{1 + e^{-(\beta_{00} + \beta_{01} x_i + \beta_{10} time_t + \beta_{11} time_t x_i + u_{0i} + u_{1i} time_t)}}. \tag{9.12}$$

The random effects u_{0i} and u_{1i} at the subject level quantify the between-subject deviation with respect to baseline score and growth rate. They are assumed normally distributed with variances σ^2_{u0} and σ^2_{u1} and covariance σ_{u01}. Note that this model does not contain a residual at the repeated measurement level. This residual is simply the discrepancy between the observed response y_{ti} and the probability p_{ti}, as follows from the model $y_{ti} = p_{ti} + e_{ti}$. The residual e_{ti} has zero mean and variance $p_{ti}(1 - p_{ti})$.

The model (9.12) assumes a linear relation between time and the logit but can be extended by including curvilinear effects. For a general introduction to multilevel models for longitudinal data with dichotomous responses the reader is referred to Hedeker and Gibbons (2006).

In Chapters 4 and 6 the mathematical expressions for the variance of the treatment effect estimator were given for continuous and dichotomous outcomes, and these expressions were found to be very similar. The expression for dichotomous outcomes was obtained from the expression for continuous outcomes by replacing σ^2_e by a function of the probabilities in both treatment conditions. For longitudinal intervention studies these probabilities vary across time and, since there are two treatment conditions, the number of probabilities is twice as large as the number of time points. For this reason, the mathematical expression for the variance of the treatment effect estimator is too complex for any practical use, especially so when the number of time periods is large.

The use of a simulation study is demonstrated on basis of a study by Visser-Van Balen et al. (2005), who studied the effect of growth hormone on the psychological functioning of adolescents with short stature. Participants were randomized to a hormone therapy condition or a control condition. Measurements were taken at baseline and once a year for three years, so the total number of measurements was four. One of the outcome variables was how the adolescents were treated by their peers; a dichotomized version of this variable is used as illustration. The categories of this variable are being treated normally versus being teased, juvenilized or ignored by peers. A relevant research question is whether the rate of change varies across the treatment conditions

Table 9.2: Results of simulation study for a longitudinal intervention study with a dichotomous outcome and 160 subjects

	θ	$\widehat{\theta}$	bias θ	sd($\widehat{\theta}$)	$\overline{se}(\widehat{\theta})$	bias se($\widehat{\theta}$)	coverage	power
β_{00}	−1.257	−1.2944	2.89	0.3496	0.3423	2.13	0.950	0.985
β_{01}	−0.109	−0.0969	12.49	0.1848	0.1831	0.93	0.950	0.086
β_{10}	1.276	1.3206	−3.38	0.4476	0.4490	−0.31	0.957	0.881
β_{11}	−0.645	−0.6690	3.59	0.2386	0.2448	−2.53	0.962	0.808
σ^2_{u0}	2.488	2.6842	−7.31	1.4098	1.3805	2.12	0.937	0.407
σ^2_{u1}	0.668	0.7131	−6.32	0.3696	0.3654	1.15	0.955	0.424
σ_{u01}	−1.103	−1.1902	−7.33	0.6525	0.6358	2.63	0.944	0.348

or, in other words, whether there is an (cross-level) interaction between time and treatment condition.

A data analysis on the basis of model (9.12) showed the p-value for a two-sided alternative is $p = 0.290$, so the interaction effect is not significant. A simulation showed that the corresponding post hoc power was only 0.235 for the sample of 36 adolescents used in the trial. The parameter estimates can be used as estimates for the true population values in a simulation study to determine the required sample size to achieve a reasonable power level of 0.8. Table 9.2 shows results of such an a priori power analysis for 160 subjects, based on 1000 generated data sets. According to the strategy of L. K. Muthén and Muthén (2002) all parameter and standard error biases should be below 10%. The table shows that only one parameter bias is slightly larger. Furthermore, all coverages of the 95% confidence intervals should be between 0.91 and 0.98, which is indeed the case. Finally the standard error of the parameter for which the power analysis is performed (i.e., β_{11} in this case) should be below 5%, and also this requirement is fulfilled. The power for the test on time by treatment interaction is 0.808, slightly above the preferred level. This is based on the assumption that there is no drop-out.

9.5 The effect of drop-out on statistical power

Thus far the focus was on trials where all subjects were measured at all and the same measurement occasions. In practice, drop-out and intermittently missed observations are omnipresent in longitudinal trials for reasons that are given in the introduction of this chapter. The main advantage of the multilevel model is that it can easily take missing data into account without discarding those subjects who have missing data at one or more measurement occasions. Including all subjects, even those with missing data, results in a higher level of power than restricting to those with complete data. This is illustrated in

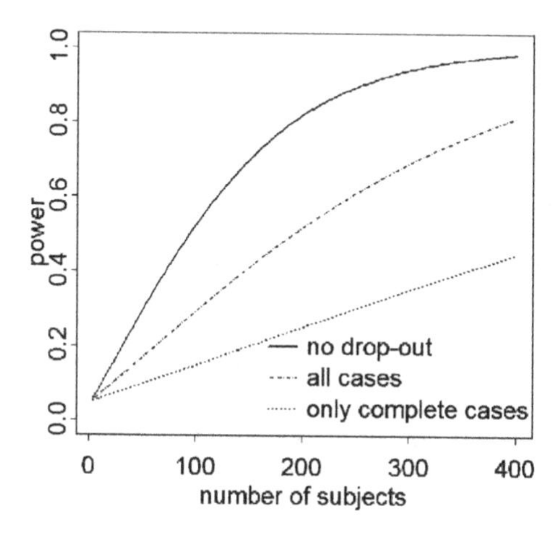

Figure 9.5: Effect of drop-out on statistical power.

Figure 9.5 that shows the power to detect a medium effect in a linear growth model with a study duration $d = 4$ and $n_1 = 5$ measurements per subject. It is assumed $\sigma_e^2/\sigma_{up}^2 = 5$, and a two-sided alternative hypothesis is used with $\alpha = 0.05$. For the case of drop-out, power levels are calculated such that 20% of the total number of subjects who started the study drop out after each measurement occasion. The figure shows the power level is highest if drop-out is absent, and in that case power is the same as for the curve with $n_1 = 5$ in the left graph of Figure 9.2. A power level of 0.8 can be achieved with $n_2 = 192$ subjects. Larger number of subjects are needed when drop-out is present. In case data from all subjects are taken into account $n_2 = 390$ subjects are needed. A sample size this large is insufficient, though, if only data from complete cases are used. In summary: drop-out decreases the power of statistical tests in longitudinal growth curves and as the multilevel model can deal with incomplete cases one is encouraged to use all data, including those of subjects who drop out during the course of the trial.

The power levels for trials with drop-out and continuous outcomes can be calculated on basis of mathematical expressions that require matrix algebra, which is outside the scope of this book. The interested reader is referred to Moerbeek (2008) for an extensive explanation how to calculate power levels for continuous outcomes. For any type of outcome variable a simulation study is a good alternative to calculate power under drop-out.

9.5.1 The effects of different drop-out patterns

As the statistical power to detect an effect of treatment depends on the drop-out pattern it is worthwhile to study and compare the effects of different drop-out patterns. One may expect power to decrease with increasing levels of drop-out and to be lower when drop-out is concentrated toward the beginning of the trial than when it is concentrated toward the end of the trial. The number of survival functions $S(t)$ that describes realistic drop-out patterns is unlimited in theory; in this section the Weibull survival function is used.

With this survival function the risk of drop-out either increases, decreases or remains constant across time. The survival function is given by $S(t) = 1 - exp(-\lambda t^{\gamma})$ with corresponding hazard function $h(t) = \lambda\gamma t^{\gamma-1}$. The shape of the survival function is determined by the drop-out parameter or hazard rate $\gamma \in [0, +\infty]$. For $\gamma < 1$ the drop-out is concentrated toward the beginning of the study, while highest drop-out rates are observed toward the end of the trial when $\gamma > 1$. Constant drop-out is observed when $\gamma = 1$ and the Weibull survival function is equal to the exponential survival function. This does, however, not necessarily imply the survival curve $S(t)$ is linear in t.

Following Galbraith and Marschner (2002) the time indicator t is rescaled by dividing by the study duration d such that $t_1 = 0$ and $t_{n_1} = 1$. In such a way t determines the proportion of the study time that has elapsed. In addition λ is replaced by $-ln(1 - \omega)$ where ln denotes the natural logarithm and $\omega \in [0, 1]$ is the proportion of subjects who drop out during the course of the study. The survival function is now rewritten as $S(t) = (1-\omega)^{t^{\gamma}}$. It should be stressed that orthogonal polynomial contrast coefficients are still used in the formulation of the multilevel model, even though the time parameter t in the survival function is rescaled.

Figure 9.6 gives survival functions for the drop-out process for three different values γ and three different values ω. The latter parameter is the proportion drop-out at duration $d = 4$. These survival functions cover a range of degrees of drop-out during the course of the trial and various points in time at which drop-out is largest.

The effect of number of subjects, number of measurement occasions and study duration for the drop-out patterns that are given in Figure 9.6 have been studied. Figures 9.7 through 9.9 show power graphs to detect a medium treatment effect in a linear growth model for a test with a two-sided alternative hypothesis and $\alpha = 0.05$. The outcome is continuous and the following values of the variance components were used to calculate the power: $\sigma^2_{u0} = \sigma_{u1} = 1$, $\sigma_{u01} = 0$ and $\sigma_e = 5$.

Figure 9.7 shows the effect of drop-out on the relation between the number of subjects and power in trials with a duration of four time units and at most five measurements per subject. The upper line corresponds to a trial without drop-out. It is observed that a lower power level is achieved when more subjects drop out during the course of the trial (i.e., when ω is larger). In addition, a comparison of the three graphs within this figure reveals that for a given drop-out proportion, a lower power level is achieved when the drop-out is concentrated toward the beginning of the trial (i.e., when $\gamma < 1$). This is obvious, since there are more people with few observations when drop-out mainly occurs at the beginning of the trial.

Figure 9.8 shows how drop-out affects the relationship between the num-

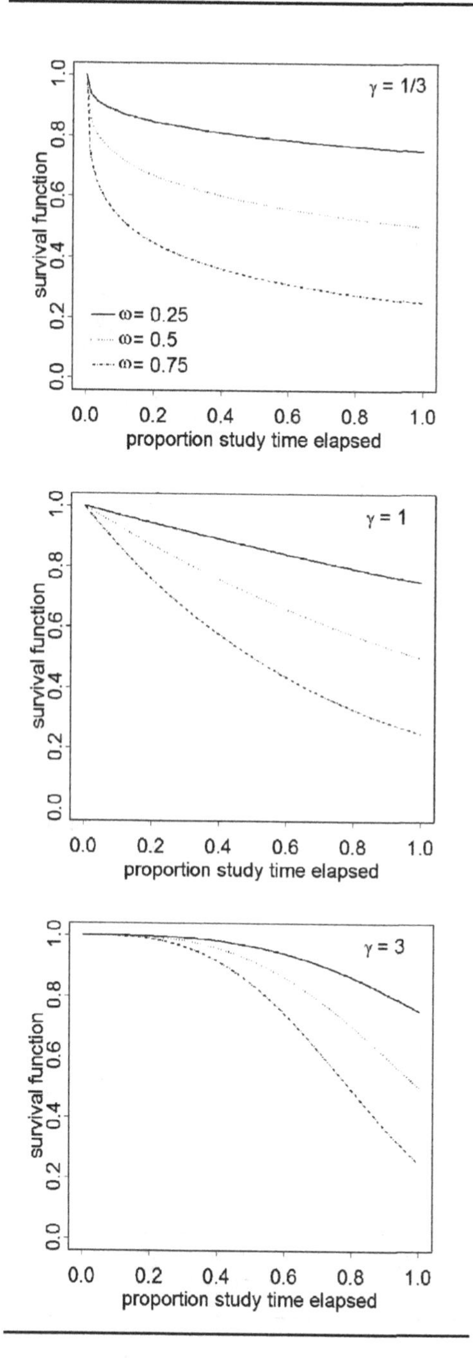

Figure 9.6: Various drop-out patterns. For legend see top panel.

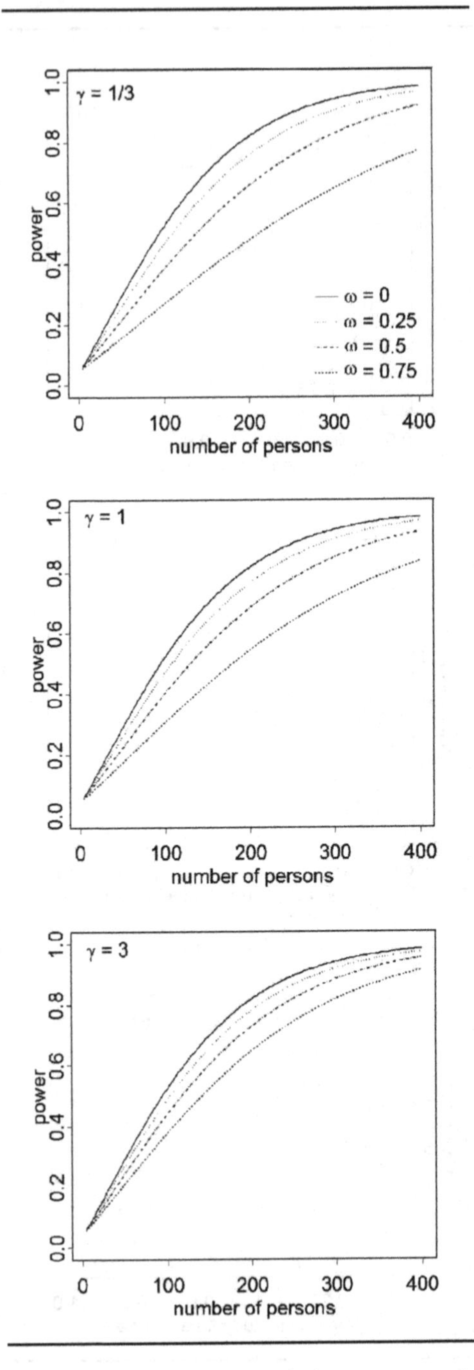

Figure 9.7: Effect of drop-out on the relation between number of subjects and power. For legend see top panel.

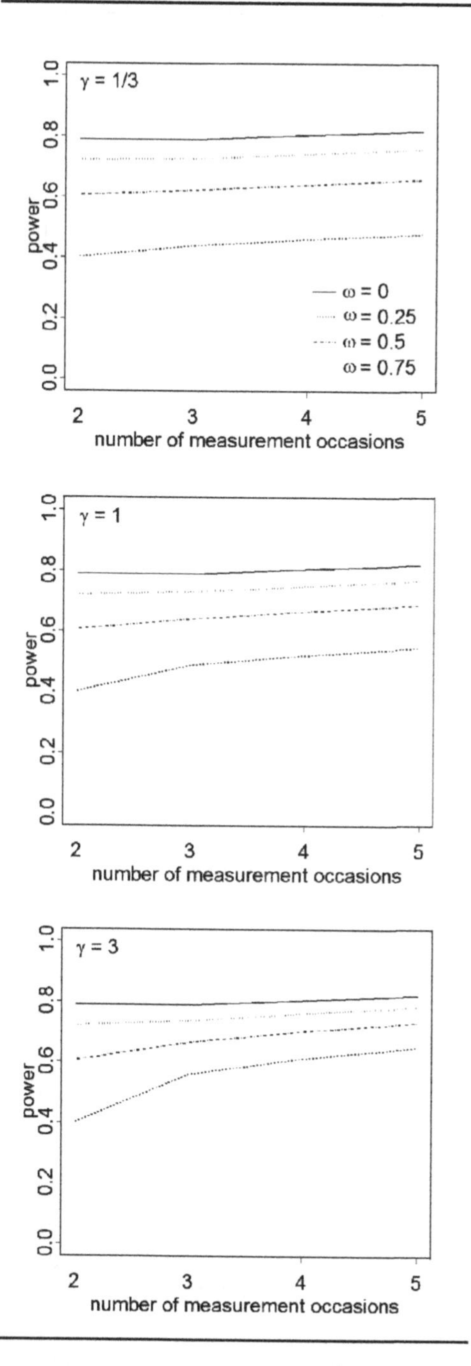

Figure 9.8: Effect of drop-out on the relation between number of measurement occasions per subject and power. For legend see top panel.

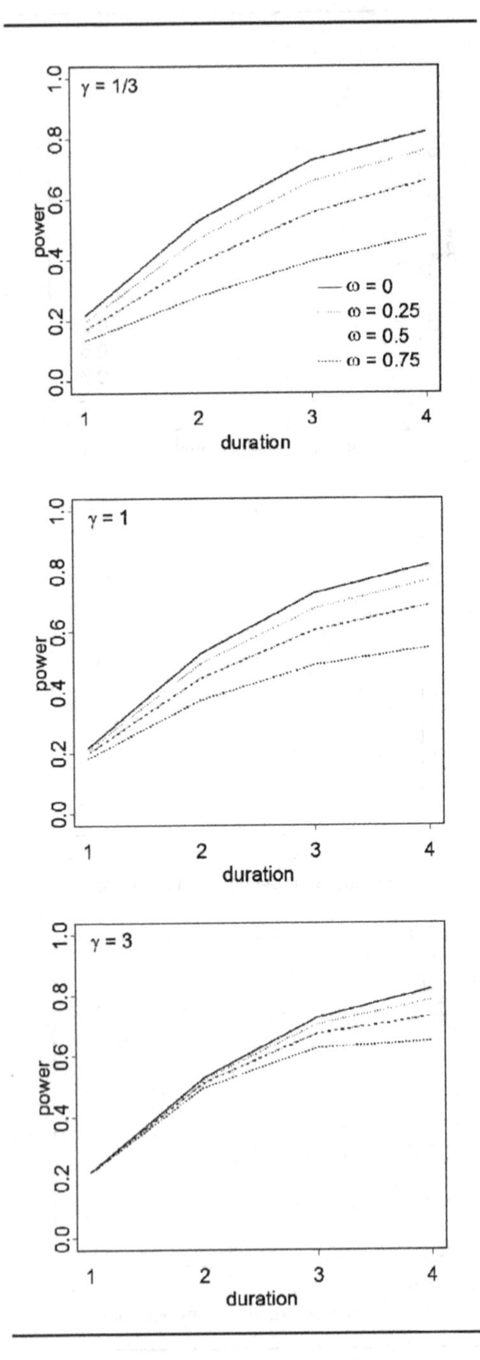

Figure 9.9: Effect of drop-out on the relation between study duration and power. For legend see top panel.

ber of measurement occasions and power for a trial with a duration of four time units and 200 subjects. Again it is observed that power decreases with increasing drop-out proportion ω and decreasing drop-out parameter γ. The drop-out parameter γ does not have an effect on power when $n_1 = 2$ since it that case there are just two measurement occasions that are located at the beginning and end of the trial. Increasing the number of measurement occasions from $n_1 = 2$ to $n_1 = 3$ has an increasing effect on power, especially when the drop-out occurs toward the end of the trial and the amount of drop-out is large. Further increasing the number of measurement occasions has a negligible effect on power.

Figure 9.9 shows how the relation between study duration and power is affected by drop-out for a trial with a maximum of $n_1 = 5$ measurements per subject and $n_2 = 200$ subjects. As in the previous two graphs drop-out has a decreasing effect on power, especially when drop-out is large and occurs toward the beginning of the trial. When drop-out becomes concentrated toward the end of the trial the difference between the power levels for various values of ω becomes smaller and for small study duration the effect of ω on power is negligible. Increasing study duration has an increasing effect on power, but this effect varies across the various drop-out patterns as determined by the values ω and γ. For instance, power hardly increases when study duration increases from $d = 3$ to $d = 4$ for a large proportion of drop-out that is mainly concentrated toward the end of the trial ($\omega = 0.75$ and $\gamma = 3$). Further increasing study duration may even result in a decrease of power. One should therefore be cautious when prolonging study duration if it will cause large amounts of drop-out.

9.5.2 Including budgetary constraints

More subjects are needed to achieve a desired power level in a trial that is hampered by drop-out than in a trial without drop-out. A naive approach would be to calculate the required number of subjects and to divide by $1 - \omega$. Such an approach assumes those who drop-out do not provide any information, even before the point when they drop out. An approach that includes all subjects, also those that drop out during the course of the trial, would be more powerful. This is illustrated in Figure 9.10 for a trial with a drop-out pattern determined by a Weibull survival function with $\omega = 0.5$ and $\gamma = 1$. That is, half of the subjects drop out during the course of the trial and they do so at a constant rate. The top two graphs in this figure correspond to the approach that only includes complete cases in the sample size calculation; the bottom two graphs use all cases, both complete and incomplete. For both approaches, it is assumed the number of measurement occasions n_1 does not exceed the value 10. The costs in the right panels are calculated relative to the costs of a design with $n_1 = 2$.

The required number of subjects for the first approach is twice as large as that in a trial without drop-out since the drop-out proportion is $\omega = 0.5$

and is corrected for by doubling the sample size. As a result, the relative costs are equal to those in a trial without drop-out as can be seen by comparing this figure to Figure 9.4. Of course, the absolute costs are twice as large as those in the trial without drop-out since the number of subjects is twice as large. Again, for a cost ratio $r = 25$ it is best to use $n_1 = 2$ measurement occasions, while for cost ratios $r = 50$ and $r = 100$ it is advised to use $n_1 = 10$ measurement occasions.

The total number of subjects needed with the second approach is lower than for the first, which can be seen by comparing the upper and lower left graphs. This, of course, results in lower trial costs. The optimal number of measurement occasions for $r = 25$ changes when the second approach is used as compared to the first: it is now advised to use $n_1 = 8$ measurement occasions. For $r = 50$ and $r = 100$ ir remains most cost-efficient to include $n_1 = 10$ measurement occasions. Of course, there will be many subjects who have fewer measurements since they drop out during the course of the study.

9.6 An example

Diabetic patients with poor glycemic control are at an increased risk of developing periodontitis. It is therefore important to develop and compare the effects of periodontal therapies for diabetic patients. Lin, Tu, Tsai, Lai, and Lu (2012) compared the effect of scaling and root planing to scaling and root planing in conjunction with antibiotics. Both treatment groups consisted of 14 patients and measurements on various inflammatory biomarkers were taken at baseline and after 3 and 6 months. Linear growth curves were fitted to the data and statistical tests showed that growth rates of the plasma concentrations did not significantly vary across treatments.

One of the reasons no significant effects were found may be the small total sample size of $n_2 = 28$. Suppose a researcher wants to replicate the trial such that it has a sufficient power of 80% to detect a difference between groups with respect to growth rate for any outcome variable. The calculation of the required number of subjects is illustrated for the outcome variable chronic reactive protein. To be able to calculate this sample size the estimates of the model parameters as found in the paper by Lin et al. (2012) are used: $\hat{\beta}_{00} = 0.227$, $\hat{\beta}_{10} = -0.107$, $\hat{\beta}_{01} = -0.032$, $\hat{\beta}_{11} = 0.031$, $\hat{\sigma}^2_{u0} = 0.89$, $\hat{\sigma}^2_{u1} = 0.032$, $\hat{\sigma}_{u01} = -0.101$ and $\hat{\sigma}^2_e = 0.2$. The parameter $\hat{\beta}_{11}$ is the estimated difference between slopes and its standardized estimate is $\hat{\delta}_1 = \hat{\beta}_{11}/\hat{\sigma}_{u1} = 0.173$, which indicates a rather small effect. The estimated variance ratio is $\sigma^2_e/\sigma^2_{u1} = 6.25$, which shows much more variation within subjects than between slopes.

The required number of subjects is calculated on basis of Equation (9.11). Given a two-sided alternative hypothesis, the required number of subjects is $n_2 = 4327.125$. This value is based on the assumption the test statistic follows

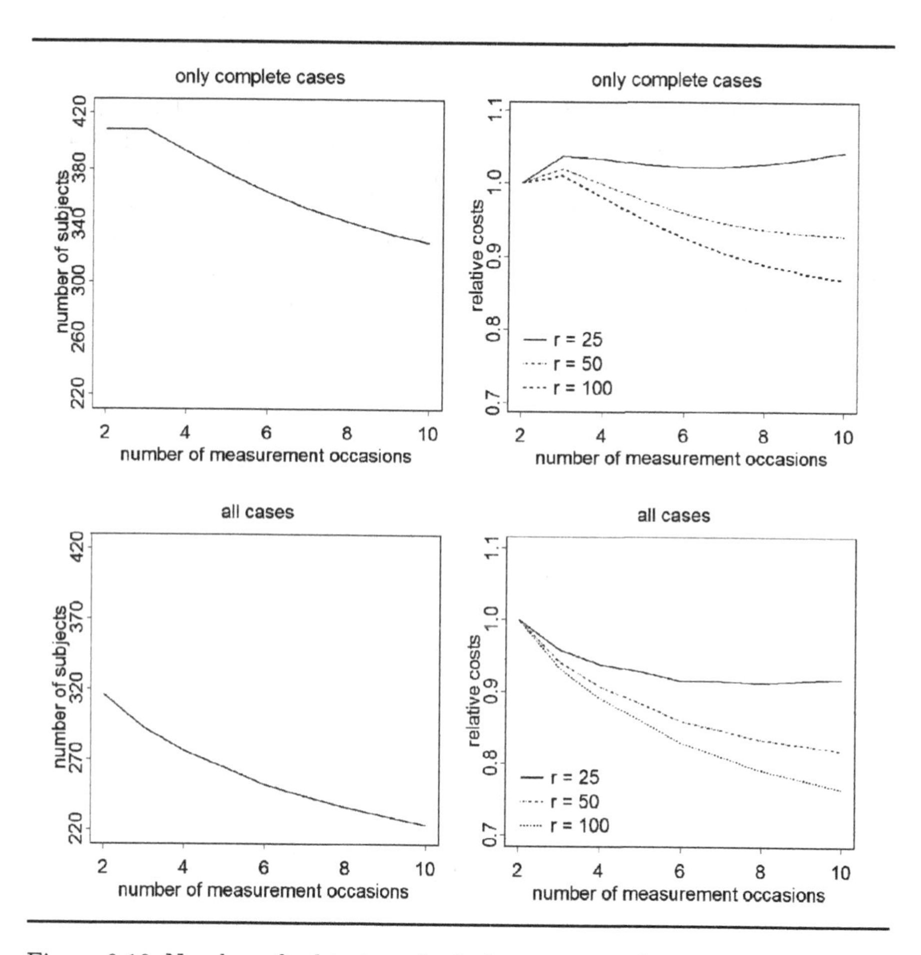

Figure 9.10: Number of subjects and relative costs as a function of the number of measurement occasions.

a normal distribution under the null hypothesis of no treatment effect. Using a t distribution with $n_2 - 2$ degrees of freedom the required sample size becomes slightly larger: $n_2 = 4330$.

It may be clear the sample size is much too large for one single experiment. A lower sample size may be achieved by decreasing the variance ratio σ_e^2/σ_{u1}^2, which can be done by decreasing the residual variance at the repeated measures level. One way to do so is by adding time-varying predictor variables that have a strong relation to the outcome. For instance, a total sample of $n_2 = 2690$ patients is needed if the variance component σ_e^2 is reduced by 50%. This is still a large value and as an alternative one could opt for replacing the effect size estimate by a larger value. Keep in mind that the effect size estimate is based on a sample of only 14 patients per treatment group and may have a large bias. For instance, a total sample of $n_2 = 1084$ patients is needed if a value twice as large as its estimate is used. This is still a large sample size but it may be possible to achieve if multiple centers participate in the trial. In that case a center level should be added to the multilevel data structure. The next chapter shows how to perform sample size calculations for three levels of nesting.

10

Extensions: three levels of nesting and factorial designs

10.1 Introduction

In the previous chapters a restriction was made to trials with two levels of nesting and two treatment conditions. In the current chapter an extension is made to three levels of nesting and factorial designs, where more than two (combinations of) treatment conditions are compared.

The first two sections focus on trials with subjects nested within clusters, which are themselves nested within sites. Examples are pupils nested within classes within schools and patients nested within therapists within clinics. If randomization to treatment conditions is done at the level of the site, all clusters and hence all subjects within a site receive the same treatment condition. Such a trial is called a three-level cluster randomized trial and is discussed in the next section. It is, however, also possible to randomize at the level of the cluster. In that case the site functions as a block and both treatments are available within each site. This allows the calculation of the treatment effect within each site and the model should account for a varying treatment effect across sites. The trial is called a multisite cluster randomized trial and is further discussed in Section 10.3.

The required sample size at each of the three levels should be calculated before the start of the trial. In other words, one should decide on the number of sites, the number of clusters per site and the number of subjects per cluster. This implies that sample size calculations become more complicated than for a trial with just two levels of nesting. Ignoring the site or cluster level while performing sample size calculations may result in an under- or overpowered study and is therefore discouraged.

The focus of Section 10.4 is on cluster randomized and multisite trials where subjects are measured repeatedly across time. In fact, these trials combine the nesting of repeated measurements within subjects (as discussed in Chapter 9) with the nesting of subjects within groups (as discussed in Chapters 4 and 6). The aim of these trials is to study subjects' changes across time, while taking into account differences across the clusters or sites in which the subjects are nested. Hence, three levels of nesting can be distinguished, for instance repeated measures within children within schools, or repeated measures within patients within therapists. For such trials it might also be tempting to ignore the highest level while calculating the required sample size, but once again this is discouraged as it may result in too high or too low sample sizes.

The focus of the final section of this chapter is on factorial designs. Consider as an example a school-based smoking prevention intervention with an in-school condition that consists of lessons on smoking and health and an out-school condition that consists of tailored letters sent to the students' homes. This design has two factors (in-school and out-school condition) with two levels each (condition absent or present). This gives a total of four combinations: control, in-school only, out-school only, both in-school and out-school, and such a design allows for the estimation of the interaction between both factors. Randomization to treatment conditions may be done at the cluster level, at the subject level, or in such a way that clusters are randomized to the levels of the one factor and subjects within clusters to the levels of the other.

10.2 Three-level cluster randomized trials

With three-level cluster randomized trials subjects are nested within clusters, which are themselves nested within sites. Randomization to treatment conditions is done at the site level so that all subjects and clusters within the same site receive the same treatment. The statistical regression model that relates a quantitative outcome y_{ijk} of subject i in cluster j in site k to treatment condition x_k is given by

$$y_{ijk} = \gamma_{000} + \gamma_{001}x_k + v_k + u_{jk} + e_{ijk}. \tag{10.1}$$

As in previous chapters, treatment condition x_k is coded -0.5 for the control condition and 0.5 for the intervention. The random effects at the site (v_k), cluster (u_{jk}) and subject (e_{ijk}) level are assumed to be independent of each other and they are assumed to follow a normal distribution with zero means and variances σ^2_v, σ^2_u and σ^2_e. The total variance of the outcome is the sum of these variance components and denoted σ^2. As treatment condition varies across sites but not across clusters within sites and not across subjects within clusters, the treatment indicator x_k has subscript k but not j and i. The sample size calculations that follow assume a balanced design with $n_3/2$ sites per treatment, n_2 clusters per site and a common cluster size n_1.

The test statistic for the test on treatment effect is calculated as $t = \hat{\gamma}_{001}/s.e.(\hat{\gamma}_{001})$ and follows a central t distribution with $n_3 - 2$ degrees of freedom if the null hypothesis of no treatment effect is true. Under the alternative hypothesis it follows a non-central t distribution with non-centrality parameter

$$\lambda = \frac{\gamma_{001}}{\sqrt{\frac{4(\sigma^2_e+n_1\sigma^2_u+n_1n_2\sigma^2_v)}{n_1n_2n_3}}}. \tag{10.2}$$

Table 10.1: The required number of sites, number of clusters per site or cluster size when the other two are known for a three-level cluster randomized trial

given	solution
n_1 and n_2	$n_3 = N\frac{1+(n_1-1)\rho_u+(n_1n_2-1)\rho_v}{n_2n_1}$
n_1 and n_3	$n_2 = N\frac{1+(n_1-1)\rho_u-\rho_v}{n_1(n_3-\rho_v N)}$
n_2 and n_3	$n_1 = N\frac{1-\rho_u-\rho_v}{(n_2n_3-\rho_u N-\rho_v n_2 N)}$

The numerator of this ratio is the unstandardized size of the treatment effect and the denominator is the standard error of its estimator (Heo & Leon, 2008; Moerbeek et al., 2000). As only one treatment per site is available the sites cannot serve as their own controls and hence the standard error is a function of the variance components at the subject, cluster and also site levels. The non-centrality parameter can be rewritten as

$$\lambda = \frac{\delta}{\sqrt{\frac{4(1+(n_1-1)\rho_u+(n_1n_2-1)\rho_v)}{n_1n_2n_3}}}, \tag{10.3}$$

where $\delta = \gamma_{001}/\sigma$ is the standardized effect size and the intracluster and intrasite correlation coefficients $\rho_u = \sigma_u^2/\sigma^2$ and $\rho_v = \sigma_v^2/\sigma^2$ measure the proportions of variance at the cluster and site levels; see also Section 2.5. The non-centrality parameter, and hence the power, increases with increasing treatment effect and sample sizes at the subject, cluster and site level and decreases with increasing intracluster and intrasite correlation coefficients. Increasing the number of sites has a higher effect on power than increasing the number of clusters per site, which in its turn has a higher effect than increasing cluster size.

In a three-level cluster randomized trial, decisions should be made on the number of sites, the number of clusters per site and the size of the clusters. In some trials two of these quantities are fixed beforehand and the third is calculated to achieve a desired power level. The requested sample sizes are given in Table 10.1. Here, $N = 2n$ is the total sample size in a trial without clustering and the sample size n per treatment condition is calculated from Equation (3.5). To facilitate getting empirical input from researchers for the sample size calculation, it can help to reformulate the sample size formulas in terms of correlations that are more easily interpreted by researchers, namely the correlation of clusters within sites and the correlation of subjects within clusters (Teerenstra, Moerbeek, Van Achterberg, Pelzer, & Borm, 2008). In terms of these, the design factor for a three-level cluster randomized trial is then the product of two design factors: one accounting for the clustering of subjects within clusters and one accounting for the clustering of clusters within sites.

When neither sample size at site, cluster or subject level is fixed beforehand a cost function can be used to calculate the optimal sample sizes at the site, cluster and subject levels (Headrick & Zumbo, 2005; Konstantopoulos, 2009, 2011; Moerbeek et al., 2000; Teerenstra et al., 2008). The total budget is denoted C and the costs per subject, per cluster and per site are denoted c_1, c_2 and c_3. The total costs at the subject level are calculated by multiplying the total number of subjects $n_1 n_2 n_3$ by the costs per subject c_1. The total costs at the cluster and site level are calculated in a similar way and the total costs are $c_1 n_1 n_2 n_3 + c_2 n_2 + c_3 n_3$. These costs should not exceed the budget C, which gives the following budgetary constraint

$$c_1 n_1 n_2 n_3 + c_2 n_2 n_3 + c_3 n_3 \leq C. \tag{10.4}$$

The sample sizes are calculated such that the standard error of the treatment effect estimator $s.e.(\hat{\gamma}_{001})$ is minimized, since this results in the largest power for the test on treatment effect given the available budget C. This gives the following optimal cluster size

$$n_1^* = \sqrt{\frac{c_2 \sigma_e^2}{c_1 \sigma_u^2}} = \sqrt{\frac{c_2(1 - \rho_u - \rho_v)}{c_1 \rho_u}}, \tag{10.5}$$

the optimal number of clusters per site

$$n_2^* = \sqrt{\frac{c_3 \sigma_u^2}{c_2 \sigma_v^2}} = \sqrt{\frac{c_3 \rho_u}{c_2 \rho_v}}, \tag{10.6}$$

and the optimal number of sites

$$\begin{aligned} n_3^* &= \frac{C}{c_3 + \sqrt{\frac{c_2 c_3 \sigma_u^2}{\sigma_v^2}} + \sqrt{\frac{c_1 c_3 \sigma_e^2}{\sigma_v^2}}} \\ &= \frac{C}{c_3 + \sqrt{\frac{c_2 c_3 \rho_u}{\rho_v}} + \sqrt{\frac{c_1 c_3 (1-\rho_u-\rho_v)}{\rho_v}}}. \end{aligned} \tag{10.7}$$

These sample sizes need to be rounded to integer values such that the budget is not exceeded and the number of sites is an even value.

Equation (10.5) shows that one would sample more subjects if the costs at the cluster level increased relative to the costs at the subject level. Also, more subjects are needed if the variance at the subject level increases relative to the variance at the cluster level. Similarly, Equation (10.6) shows that the optimal number of clusters increases if the costs at the site level increase or the costs at the cluster level decrease. More clusters are needed when the cluster level variance increases relative to the site level variance. Finally, from Equation (10.7) it follows that the optimal number of sites increases as the budget C increases. The other two sample sizes do not depend on the budget.

Given the optimal sample sizes, the standard error of the treatment effect estimator is

$$
\begin{aligned}
s.e.(\hat{\gamma}_{001}) &= 2\frac{\sqrt{c_1\sigma_e^2}+\sqrt{c_2\sigma_u^2}+\sqrt{c_3\sigma_v^2}}{\sqrt{C}} \\
&= 2\sigma\frac{\sqrt{c_1(1-\rho_u-\rho_v)}+\sqrt{c_2\rho_u}+\sqrt{c_3\rho_v}}{\sqrt{C}}. \qquad (10.8)
\end{aligned}
$$

In practice this value will be somewhat larger due to rounding of the optimal sample sizes to integer values. As is obvious, this standard error decreases with increasing budget C since a higher budget means a larger sample can be taken.

Instead of maximizing the power given a fixed budget C it is also possible to minimize the budget given a fixed power level $1-\beta$. For this strategy the optimal sample sizes are also given by Equations (10.5) to (10.7). The required budget is equal to

$$
C = 4\frac{(\sqrt{c_1(1-\rho_u-\rho_v)}+\sqrt{c_2\rho_u}+\sqrt{c_3\rho_v})^2(z_{1-\alpha}+z_{1-\beta})^2}{\delta^2}, \qquad (10.9)
$$

and will become a little larger if sample sizes are rounded to integer values. This equation holds for a test with a one-sided alternative hypothesis; for a two-sided alternative hypothesis the type I error rate α should be replaced by $\alpha/2$. This equation should be applied when the normal approximation to the t distribution is used, which is only justified when the $df = n_3 - 2$ are large. Using the t distribution, the degrees of freedom depend on the number of clusters and the optimal design does not necessarily follow from Equations (10.5) to (10.7). The optimal design can then be found by evaluating a large number of designs (n_1, n_2, n_3) and selecting the one for which the desired power level is achieved at lowest cost. It may be clear that this approach may be very time consuming.

Figure 10.1 shows the optimal design for a three-level cluster randomized trial. This figure depicts contour lines of the $s.e.(\hat{\beta}_1)$ in the (n_1, n_2) plane for a hypothetical example with variance components $\sigma_e^2 = 160$, $\sigma_u^2 = 20$, $\sigma_v^2 = 5$. These values correspond to 2.7% variance at the site level, 10.8% variance at the cluster level and 86.5% variance at the subject level, and agree with the fact that the proportion variance at the subject level is often much larger than that at the cluster level, which in its turn is often much larger than that at the site level. The costs are $c_1 = 100$, $c_2 = 200$ and $c_3 = 300$ and the budget is fixed at $C = 200,000$. Here a realistic situation with higher costs at higher levels of the multilevel data structure is observed.

The curved dashed line represents designs with $n_3 = 20$ and all designs below this line have $n_3 > 20$. A minimum of 20 sites is assumed since a sample with fewer sites can hardly be regarded as representative of the population it is drawn from. The vertical and horizontal dashed lines correspond to designs with $n_1 = 2$ and $n_2 = 2$, respectively. To preserve the three-level data structure

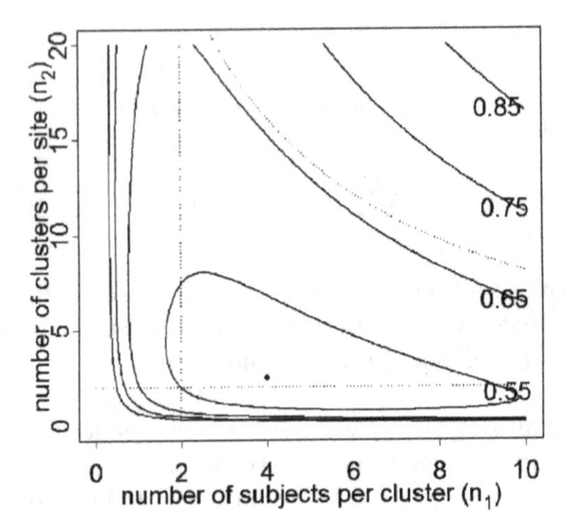

Figure 10.1: Graphical representation of $s.e.(\hat{\beta}_1)$ for a three-level cluster randomized trial as a function of n_1 and n_2. The collection of feasible designs is enclosed in the dashed lines.

the focus is on designs with at least two clusters per site and at least two subjects per cluster. All feasible designs can be found in the region that is enclosed by the dashed lines. The optimal design is found in the feasible design region at $n_1 = 4$, $n_2 = 2.5$, $n_3 = 134.3$ and it is depicted by a dot. As the optimal values n_2 and n_3 are not integer values they have to be rounded. This implies the $s.e.(\hat{\beta}_1)$ will increase somewhat.

The figure can also be used to study the efficiency of a design that differs from the optimal design. For this specific set of variance components and costs, changing the number of subjects per cluster has a higher effect on design efficiency than changing the number of clusters per site by the same amount.

10.3 Multisite cluster randomized trials

With multisite cluster randomized trials randomization to treatment conditions is performed at the cluster level within each site. As both treatment conditions are available within each site, the site functions as its own control. Therefore, a higher power level can be achieved than with a three-level cluster randomized trial, provided contamination of the control group is absent. Also, it allows us to study whether the effect of treatment varies across sites. In the calculations that follow a balanced design is assumed with n_1 subjects per cluster, n_2 clusters per site and the number of sites is denoted

n_3. Within each site $n_2/2$ clusters are randomized to the control and the other $n_2/2$ clusters are randomized to the intervention.

Our first focus is on models where the effect of treatment does not vary across sites. The subject level model for subject i in cluster j in site k is given by

$$y_{ijk} = \pi_{0j} + e_{ijk}. \tag{10.10}$$

The mean scores π_{0j} vary across the clusters j and are modeled as a function of treatment condition x_{jk}:

$$\pi_{0j} = \beta_{00k} + \beta_{01k}x_{jk} + u_{jk}. \tag{10.11}$$

Treatment condition is coded -0.5 and 0.5 for the intervention and control so that β_{00k} is the mean score in site k and β_{01k} is the effect of intervention in site j. The first but not the latter is assumed to vary randomly across sites:

$$\begin{aligned} \beta_{00k} &= \gamma_{000} + v_k \\ \beta_{01k} &= \gamma_{001}. \end{aligned} \tag{10.12}$$

The related single equation model is now equal to

$$y_{ijk} = \gamma_{000} + \gamma_{010}x_{jk} + v_k + u_{jk} + e_{ijk}. \tag{10.13}$$

The main difference between this model and the model for a three-level cluster randomized trial in Equation (10.1) is the subscript for treatment condition x. This subscript is jk for a multisite cluster randomized trial as treatment varies across clusters within sites. The related regression weight γ_{010} is the effect of treatment. The error terms e_{ijk}, u_{jk} and v_k represent sources of unexplained variability at the subject, cluster and site levels. They are assumed to be independent of each other and to follow a normal distribution with zero mean and variances σ_e^2, σ_u^2 and σ_v^2. The total sum of these variance components is denoted σ^2.

The test statistic $t = \hat{\gamma}_{010}/s.e.(\hat{\gamma}_{010})$ is used to test the effect of treatment and under the null hypothesis of no treatment effect it follows a central t distribution with $n_2(n_3-1)$ degrees of freedom. Under the alternative hypothesis it follows a non-central t distribution with non-centrality parameter

$$\lambda = \frac{\gamma_{010}}{\sqrt{\frac{4(\sigma_e^2+n_1\sigma_u^2)}{n_1 n_2 n_3}}} = \frac{\delta}{\sqrt{\frac{4(1+(n_1-1)\rho_u)}{n_1 n_2 n_3}}}, \tag{10.14}$$

where $\delta = \gamma_{010}/\sigma$ is the standardized effect size. The numerator is the treatment effect, the denominator the standard error of its estimator. This non-centrality parameter decreases when the treatment effect increases, when the variance components decrease and when the sample sizes at the subject, cluster and site levels increase. Increasing the number of sites has the same effect as increasing the number of clusters per site, but the effect of increasing the number of subjects per cluster is smaller.

Table 10.2: The required number of sites, number of clusters per site or cluster size when the other two are known for a multisite cluster randomized trial with a fixed effect of treatment

given	solution
n_1 and n_2	$n_3 = N\frac{1+(n_1-1)\rho_u}{n_2 n_1}$
n_1 and n_3	$n_2 = N\frac{1+(n_1-1)\rho_u}{n_1 n_3}$
n_2 and n_3	$n_1 = N\frac{1-\rho_u}{(n_2 n_3-\rho_u N)}$

The $s.e.(\hat{\gamma}_{010})$ does not depend on the between-site variance σ_v^2, so randomization of clusters within sites results in a higher power to detect an effect of treatment than randomization of sites. This implies the required sample sizes to achieve a desired power level will also be lower. These sample sizes are given in Table 10.2 and are derived from those in Table 10.1 by setting $\rho_v = 0$. Again, $N = 2n$ is the total number of subjects in a trial without nesting and n is the number of subjects per condition.

When neither of the three sample sizes n_1, n_2 and n_3 is fixed beforehand one can take budgetary constraints into account and the cost function (10.4) can be used to calculate the optimal design. The optimal number of subjects per cluster is identical as for a three-level cluster randomized trial and is given by (10.5). The sample sizes at the other two levels differ from those of a three-level cluster randomized trial. The optimal number of clusters per site is

$$n_2^* = \frac{C - 20c_3}{20\sqrt{\frac{c_1 c_2 \sigma_e^2}{\sigma_u^2}} + 20c_2}, \qquad (10.15)$$

given that the number of sites is fixed at $n_3 = 20$ so that the sample of sites provides enough information on the population it is drawn from. The number of clusters per site increases as the budget increases, while the other two sample sizes do not depend on the trial budget. In the previous section it was observed for three-level cluster randomized trials that the number of sites is the only sample size that depends on the budget. We can conclude that the budget determines the sample size at the level at which randomization to treatments is done.

When the optimal sample sizes are used, the standard error of the treatment effect estimator is

$$s.e.(\hat{\gamma}_{001}) = 2\frac{\sqrt{c_1\sigma_e^2} + \sqrt{c_2\sigma_u^2}}{\sqrt{C - 20c_3}} = 2\sigma\frac{\sqrt{c_1(1-\rho_u-\rho_v)} + \sqrt{c_2\rho_u}}{\sqrt{C - 20c_3}}. \qquad (10.16)$$

Rounding the sample sizes to integer values results in a somewhat larger standard error and lower power. As is obvious, this standard error decreases with increasing budget C.

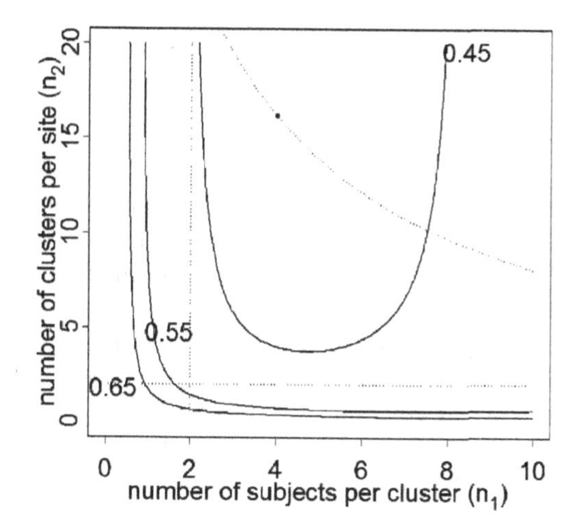

Figure 10.2: Graphical representation of $s.e.(\hat{\beta}_1)$ for a multisite cluster randomized trial as a function of n_1 and n_2. The collection of feasible designs is enclosed in the dashed lines.

Given the optimal sample sizes, the budget to achieve a power $1-\beta$ in a test with a one-sided alternative hypothesis is equal to

$$C = 4\frac{(\sqrt{c_1(1-\rho_u-\rho_v)}+\sqrt{c_2\rho_u})^2(z_{1-\alpha}+z_{1-\beta})^2}{\delta^2} - 20c_3, \qquad (10.17)$$

and for a two-sided alternative hypothesis α is replaced by $\alpha/2$. This approach relies on the normal approximation to the t distribution and works well when the *df* are large. If the t distribution is used, the optimal design is found by evaluating a large number of designs (n_1, n_2, n_3) and selecting the one for which the desired power level is achieved at lowest costs.

Figure 10.2 shows the optimal design for a multisite cluster randomized trial with variance components $\sigma_e^2 = 160$, $\sigma_u^2 = 20$, $\sigma_v^2 = 5$. The costs are $c_1 = 100$, $c_2 = 200$ and $c_3 = 300$ and the budget is fixed at $C = 200,000$. These are the same values as in the example in the previous section for a three-level cluster randomized trial. The optimal design is found at the border of the feasible design region at $n_1 = 4$, $n_2 = 16.2$ and $n_3 = 20$ and is depicted by a dot in the figure. A comparison of Figures 10.1 and 10.2 shows that for any combination (n_1, n_2, n_3) lower standard errors are achieved with a multisite cluster randomized trial. Furthermore, the shape of the contour lines for a multisite cluster randomized trial differs from the shape for a three-level cluster randomized trial and this explains the difference in the location of the optimal design.

Varying treatment effect. The assumption of a constant treatment effect across sites may not always be realistic. If the effect of treatment is likely to vary across sites it should be accounted for in the statistical model. The model for subject i in cluster j in site k is represented as

$$y_{ijk} = \gamma_{000} + \gamma_{010}x_{jk} + v_{0k} + v_{1k}x_{jk} + u_{jk} + e_{ijk}. \tag{10.18}$$

The difference between this model and the model that assumes a fixed effect of treatment across sites is the addition of the interaction term $v_{1k}x_{jk}$. This is a product of the random term v_{1k} and the treatment indicator x_{ij}. The treatment effect does not vary across clusters within sites if the variance σ^2_{v1} of the random term v_{1k} is equal to zero. The site level variances σ^2_{v0} and σ^2_{v1} measure the between-site variability with respect to the mean outcome and treatment effect. Their covariance is denoted σ_{v01}. As in Chapter 6, this covariance is set equal to zero to calculate the optimal design. First, note that the total outcome variance is

$$var(y_{ijk}) = \sigma^2 = \sigma^2_{v0} + \frac{1}{4}\sigma^2_{v1} + \sigma^2_u + \sigma^2_e. \tag{10.19}$$

The proportions of variance at the cluster and site levels are $\rho_u = \sigma^2_u/(\sigma^2_{v0} + \frac{1}{4}\sigma^2_{v1} + \sigma^2_u + \sigma^2_e)$ and $\rho_v = (\sigma^2_{v0} + \frac{1}{4}\sigma^2_{v1})/(\sigma^2_{v0} + \frac{1}{4}\sigma^2_{v1} + \sigma^2_u + \sigma^2_e)$, respectively. The latter may be split in a part ρ_{v0} due to the random intercept and another part ρ_{v1} due to the random slope: $\rho_{v0} = (\sigma^2_{v0})/(\sigma^2_{v0} + \frac{1}{4}\sigma^2_{v1} + \sigma^2_u + \sigma^2_e)$ and $\rho_{v1} = (\frac{1}{4}\sigma^2_{v1})/(\sigma^2_{v0} + \frac{1}{4}\sigma^2_{v1} + \sigma^2_u + \sigma^2_e)$.

The average treatment effect is denoted γ_{010} and its significance is tested by the test statistic $t = \hat{\gamma}_{010}/s.e.(\hat{\gamma}_{010})$. Under the null hypothesis of no treatment effect it follows a central t distribution with $n_3 - 1$ degrees of freedom, while under the alternative hypothesis the non-centrality parameter of the non-central t distribution is

$$\lambda = \frac{\gamma_{010}}{\sqrt{\frac{4(\sigma^2_e + n_1\sigma^2_u + \frac{1}{4}n_1n_2\sigma^2_{v1})}{n_1n_2n_3}}} = \frac{\delta}{\sqrt{\frac{4(1-\rho_{v0}+(n_1-1)\rho_u+(n_1n_2-1)\rho_{v1})}{n_1n_2n_3}}}. \tag{10.20}$$

The numerator of the middle term is the treatment effect while the denominator is the standard error of its estimator. As in previous sections, $\delta = \gamma_{010}/\sigma$ is the standardized effect of treatment and the right term shows the non-centrality parameter as a function of the standardized treatment effect and the variance proportions ρ_u, ρ_{v0} and ρ_{v1}.

Table 10.3 gives the optimal number of sites, clusters per site and cluster size when the other two are fixed beforehand. As previously, N is the total sample size in a trial without nesting.

In some trials none of the samples is fixed beforehand. The optimal design is the one that minimizes $var(\hat{\gamma}_{010})$ subject to the budgetary constraint (10.4). The optimal number of subjects per cluster is equal to

$$n_1^* = \sqrt{\frac{c_2\sigma^2_e}{c_1\sigma^2_u}} = \sqrt{\frac{c_2(1-\rho_u-\rho_{v0}-\rho_{v1})}{c_1\rho_u}}, \tag{10.21}$$

Table 10.3: The required number of sites, clusters per site or cluster size when the other two are known for a multisite cluster randomized trial with a varying effect of treatment

given	solution
n_1 and n_2	$n_3 = N\frac{1-\rho_{v0}+(n_1-1)\rho_u+(n_1n_2-1)\rho_{v1}}{n_2n_1}$
n_1 and n_3	$n_2 = N\frac{1-\rho_{v0}+(n_1-1)\rho_u-\rho_{v1}}{n_1(n_3-\rho_{v1}N)}$
n_2 and n_3	$n_1 = N\frac{1-\rho_{v0}-\rho_u-\rho_{v1}}{(n_2n_3-\rho_u N-\rho_{v1}n_2N)}$

the optimal number of clusters per site is

$$n_2^* = \sqrt{\frac{c_3\sigma_u^2}{c_2\frac{1}{4}\sigma_{v1}^2}} = \sqrt{\frac{c_3\rho_u}{c_2\rho_{v1}}}, \tag{10.22}$$

and the optimal number of sites is

$$\begin{aligned} n_3^* &= \frac{C}{c_3 + \sqrt{\frac{c_2c_3\sigma_u^2}{\frac{1}{4}\sigma_{v1}^2}} + \sqrt{\frac{c_1c_3\sigma_e^2}{\frac{1}{4}\sigma_{v1}^2}}} \\ &= \frac{C}{c_3 + \sqrt{\frac{c_2c_3\rho_u}{\rho_{v1}}} + \sqrt{\frac{c_1c_3(1-\rho_u-\rho_{v0}-\rho_{v1})}{\rho_{v1}}}}. \end{aligned} \tag{10.23}$$

The difference between the optimal sample size equations for a cluster randomized trial and those for a multisite trial with varying treatment effect across sites is that those for a cluster randomized trial include the term σ_v^2 while those for a multisite trial include the term $\frac{1}{4}\sigma_{v1}^2$.

10.4 Repeated measures in cluster randomized trials and multisite trials

The focus of Chapter 9 was on longitudinal intervention studies where subjects are randomly assigned to treatment conditions and measured at multiple time periods during the course of the trial. Multilevel models are used to allow for between-subject variability with respect to initial status and growth. The main question in a longitudinal trial is whether treatment condition has an effect on polynomial growth parameters. In linear growth models, for instance, the question is whether the constant growth across time varies across treatment conditions.

The aim of this section is to study optimal designs for longitudinal studies where subjects are nested within clusters. An example is a school-based smoking prevention intervention with randomization to treatment conditions at the school level and multilevel measurements across time on each subject. This is a cluster randomized trial as randomization to treatment conditions is done at the highest level. An example of a multisite trial with repeated measures is a trial with patients who suffer from morbid obesity and are nested within clinics. Randomization to treatment conditions, such as gastric bypass or diet and lifestyle changes, is done at the patient level and measurements on patients' health is done at multiple points across time. As both treatment conditions are available within each clinic, this design allows us to study whether rates of change vary across clinics. The sample size equations in this section are based on the findings in De Jong, Moerbeek, and Van Der Leeden (2010) and Heo and Leon (2009). For simplicity, a restriction is made to linear growth models and all subjects are measured at the same measurement occasions. Equidistant points in time are used, which implies the frequency of measurement is constant across time.

The first focus is on studies with randomization at the cluster level. The number of measurements per subject is denoted n_1, the number of subjects per cluster is n_2 and the number of clusters is n_3. Equal randomization to the intervention and control conditions is assumed. The model at time point t within subject i in cluster j is given by the equation

$$y_{tij} = \pi_{0ij} + \pi_{1ij} time_{tij} + e_{tij}. \quad (10.24)$$

Here, π_{0ij} and π_{1ij} are the baseline measurement (i.e., intercept) and rate of change (i.e., slope) that are assumed to vary across subjects and clusters. The error terms e_{tij} represent the discrepancy between the observed score y_{tij} and the score as predicted from the regression equation. These errors are assumed to be independently and normally distributed with zero mean and variance σ^2_e.

A model is formulated at the subject level to model inter-subject variability with respect to intercept and slope

$$\begin{aligned} \pi_{0ij} &= \beta_{00j} + u_{0ij} \\ \pi_{1ij} &= \beta_{10j} + u_{1ij}. \end{aligned} \quad (10.25)$$

Here, all variability is assumed random but of course predictor variables can be added to the model to explain part of the variability. The regression weights β_{00j} and β_{01j} are the mean intercept and slope within cluster j and the error terms u_{0ij} and u_{1ij} represent the discrepancies of subject i in cluster j from these mean scores. It is assumed these errors follow a bivariate normal distribution with zero mean and variances σ^2_{u0} and σ^2_{u1}, and their covariance is denoted σ_{u01}.

The between-cluster variability is modeled using treatment condition x_j as explanatory variable:

$$\begin{aligned} \beta_{00j} &= \gamma_{000} + \gamma_{001} x_j + v_{0j} \\ \beta_{10j} &= \gamma_{100} + \gamma_{101} x_j + v_{1j}. \end{aligned} \quad (10.26)$$

Treatment condition is coded $x_j = -0.5$ for the control group and $x_j = 0.5$ for the intervention so that γ_{000} and γ_{100} are the mean intercept and slope and γ_{001} and γ_{101} are the effects of treatment on the intercept and slope. The error terms v_{0j} and v_{1j} are assumed to follow a bivariate normal distribution with a zero mean vector, variances σ^2_{v0} and σ^2_{v1}, and covariance σ_{v01}.

A combination of Equations (10.24), (10.25) and (10.26) gives the composite model

$$\begin{aligned} y_{tij} &= \gamma_{000} + \gamma_{100} time_{tij} + \gamma_{001} x_j + \gamma_{101} time_{tij} x_j \\ &\quad + v_{0j} + v_{1j} time_{tij} + u_{0ij} + u_{1ij} time_{tij} + e_{tij}. \end{aligned} \tag{10.27}$$

The first line at the right side is the fixed part of the model and this is the expected outcome for the measure at time point t in subject i in cluster j. Note that the fixed part has a cross-level interaction term $\gamma_{101} time_{tij} x_j$ that allows the effect of time to vary across treatment conditions. In other words, it specifies to what extent the linear change (growth or decline) across time varies across treatment conditions. To test whether the parameter γ_{101} is significant, a t test is performed with a test statistic $t = \hat{\gamma}_{101}/s.e.(\hat{\gamma}_{101})$ that follows a central t distribution with $n_3 - 2$ degrees of freedom under the null hypothesis of no differential growth. Under the alternative it follows a non-central t distribution with non-centrality parameter

$$\lambda = \frac{\gamma_{101}}{\sqrt{\frac{4(\sigma^2_e + s^2 n_1 \sigma^2_{u1} + s^2 n_1 n_2 \sigma^2_{v1})}{s^2 n_1 n_2 n_3}}}, \tag{10.28}$$

where

$$s^2 = \frac{\sum_{t=1}^{n_1} (time_{tij} - \overline{time})^2}{n_1} \tag{10.29}$$

is the biased variance of the time points. The standard error in the denominator of Equation (10.28) is a function of this biased variance, the sample sizes at each level of the hierarchical data structure, the error variance at the repeated measures level, and between-subject and between-cluster variability of the linear growth. The between-subject and between-cluster variabilities of the baseline scores are not part of this standard error.

When two of the sample sizes n_1, n_2 and n_3 are fixed beforehand, the third can be calculated such that a desired power level $1 - \beta$ is achieved for testing the time by treatment interaction. As the degrees of freedom of the t distribution depend on the number of clusters n_3, it is not possible to derive an explicit equation for this design factor when the other two are fixed. For this reason, the t distribution is approximated by the standard normal and this approximation works well when the number of clusters is large. The required number of clusters is then given by

$$n_3 = 4\frac{\sigma^2_e + s^2 n_1 \sigma^2_{u1} + s^2 n_1 n_2 \sigma^2_{v1}}{s^2 n_1 n_2} \left(\frac{z_{1-\alpha} + z_{1-\beta}}{\gamma_{101}} \right)^2 \tag{10.30}$$

for a one-sided test with type I error α (replace α by $\alpha/2$ for a two-sided test). In the same manner, an equation for the required number of subjects per cluster can be derived if the number of clusters and the number of measurements per cluster are given:

$$n_2 = 4\frac{\sigma_e^2 + s^2 n_1 \sigma_{u1}^2}{s^2 n_1 n_3 \left(\left(\frac{\gamma_{101}}{z_{1-\alpha}+z_{1-\beta}}\right)^2 - 4\frac{\sigma_{v1}^2}{n_3}\right)}. \tag{10.31}$$

Unfortunately, it is not possible to derive an explicit equation for the required number of measurements per subject in a trial where the number of subjects per cluster and the number of clusters is fixed. This can be explained by the fact that the variance s^2 is a function of n_1 and the study duration. To calculate the required n_1, one can calculate the power for a large range of values n_1 and select the smallest value n_1 that results in the desired power level. Furthermore, as the variance s^2 depends on n_1 it is not possible to derive the optimal sample sizes such that the variance of the treatment effect estimator is minimized and the budget for sampling clusters and subjects and taking repeat measurements is not exceeded. It is, of course, possible to compare the costs of various designs that have the same level of power and to select the least expensive design. Section 9.3.3 shows how this can be done for longitudinal trials where there is no nesting of subjects within clusters or sites and the same approach can be applied for cluster randomized and multisite trials with repeated measurements.

Now that designs with repeated measures in cluster randomized trials have been discussed, it is time to continue with multisite trials with repeated measurements. With such trials, randomization is done at the subject level within each site and both treatment conditions are available in each site. Again a balanced design is assumed with n_3 sites, n_2 subjects per site and n_1 measurements per subject with a fixed frequency of measurement during the course of the study. Within each site, $n_2/2$ subjects are randomized to the intervention condition and the others to the control.

The model at the repeated measures level is the same as for a cluster randomized trial:

$$y_{tij} = \pi_{0ij} + \pi_{1ij} time_{tij} + e_{tij}, \tag{10.32}$$

where e_{tij} is a random error at the repeated measures level that is assumed to follow a normal distribution with zero mean and variance σ_e^2. The random intercept π_{0ij} and slope π_{1ij} are now predicted by the treatment condition x_{ij}:

$$\begin{aligned} \pi_{0ij} &= \beta_{00j} + \beta_{01j} x_{ij} + u_{0ij} \\ \pi_{1ij} &= \beta_{10j} + \beta_{11j} x_{ij} + u_{1ij}. \end{aligned} \tag{10.33}$$

The random errors u_{0ij} and u_{1ij} are assumed to follow a bivariate normal distribution with zero means, variances σ_{u0}^2 and σ_{u1}^2 and covariance σ_{u01}^2. This model has four random effects: a random intercept β_{00j}, a random effect of time β_{10j}, a random effect of treatment β_{01j} and a random time by treatment interaction β_{11j}. The model at the site level is as follows:

$$\begin{aligned}
\beta_{00j} &= \gamma_{000} + v_{0j} \\
\beta_{10j} &= \gamma_{100} + v_{1j} \quad (10.34)\\
\beta_{01j} &= \gamma_{010} + v_{2j} \quad (10.35)\\
\beta_{11j} &= \gamma_{110} + v_{3j}, \quad (10.36)
\end{aligned}$$

and can be extended by adding predictor variables at the site level. The vector of random effects $\mathbf{v}$ at the site level is assumed to follow a multivariate normal distribution with a zero mean vector and covariance matrix

$$cov(\mathbf{v}) = \begin{pmatrix} \sigma^2_{v0} & \sigma_{v01} & \sigma_{v02} & \sigma_{v03} \\ \sigma_{v01} & \sigma^2_{v1} & \sigma_{v12} & \sigma_{v13} \\ \sigma_{v02} & \sigma_{v12} & \sigma^2_{v2} & \sigma_{v23} \\ \sigma_{v03} & \sigma_{v13} & \sigma_{v23} & \sigma^2_{v3}. \end{pmatrix}. \quad (10.37)$$

The multilevel model can be presented using a single equation

$$\begin{aligned}
y_{tij} = {} & \gamma_{000} + \gamma_{100} time_{tij} + \gamma_{010} x_{ij} + \gamma_{110} time_{tij} x_{ij} \\
& + v_{0j} + v_{1j} time_{tij} + v_{2j} x_{ij} + v_{3j} time_{tij} x_{ij} \quad (10.38)\\
& + u_{0ij} + u_{1ij} time_{tij} + e_{tij}. \quad (10.39)
\end{aligned}$$

The first line at the right side is the fixed part of the model; the second and third lines represent the random parts.

The fixed parameter γ_{110} is the average time by treatment interaction across sites. In other words, it is the average effect of treatment condition on the linear change across time. The significance of this effect can be tested using the test statistic $t = \hat{\gamma}_{110}/s.e.(\hat{\gamma}_{110})$ which follows a central t distribution with $n_3 - 1$ degrees of freedom under the null hypothesis of no treatment effect. Under the alternative hypothesis of non-zero treatment effect, it follows a non-central t distribution with non-centrality parameter

$$\lambda = \frac{\gamma_{110}}{\sqrt{\frac{4(\sigma^2_e + s^2 n_1 \sigma^2_{u1} + \frac{1}{4} s^2 n_1 n_2 \sigma^2_{v3})}{s^2 n_1 n_2 n_3}}}, \quad (10.40)$$

where s^2 is the biased variance of the time points as given by (10.29). The numerator of the non-centrality parameter is the effect of treatment on linear change, while the denominator is the standard error of its estimator. A comparison of this equation with the non-centrality parameter for a cluster randomized trial as given by (10.28) reveals that the only difference in the standard errors for these two designs is that the variance component σ^2_{v1} in (10.28) is replaced by $\frac{1}{4}\sigma^2_{v3}$ in (10.40). This implies the required number of sites to achieve a power $1 - \beta$ for testing the significance of γ_{110} follows from (10.30) by replacing σ^2_{v1} by $\frac{1}{4}\sigma^2_{v3}$ and γ_{101} by γ_{110}. Making the same replacements, the required number of subjects per site is found from (10.31).

10.5 Factorial designs

In a factorial design with two factors, subjects and/or clusters are allocated to each combination of the two independent factors. For example, a Dutch smoking prevention study (Ausems et al., 2002) investigated the effect of an in-school intervention (a seven-lesson program during school classes) and out-of-school intervention (tailored letters to the students' homes). Classes were randomized to either no intervention, only in-school intervention, only out-of-school intervention, or the combination of in-school and out-of-school intervention. In general, a factorial design may be worthwhile when two or more interventions, or more generally factors, can be combined and the interest is in the effect of the interventions separately but also in combinations.

This was the case in the above study, because the out-of-school intervention was expected to enhance the in-school intervention, and therefore interest was not only in the comparing the two interventions, but also in the effect of combining these two interventions. In a factorial design, the factors of interest need not always be interventions. One or more factors may be observational variables, either on subject level such as gender or on cluster level such as type of school.

As the dependent variables (e.g., the effect measures) may be continuous or binary, these will be discussed in separate subsections below. The factors will be considered dichotomous and one quarter of all subjects ($\frac{1}{4}n_2n_1$) will be allocated to each of the $4 = 2 \times 2$ combinations of factors. This may seem restrictive, but Lemme, Van Breukelen, and Berger (in press) showed that for a range of research questions this is an optimal or a highly efficient allocation of clusters in case of a continuous outcome.

10.5.1 Continuous outcome

Here the focus is on models without interaction between the two factors; see Moerbeek et al. (2003c) for the case with interaction. In a factorial design with two factors x_{1ij} and x_{2ij}, the outcome y_{ij} for subject i in cluster j can be modeled as

$$y_{ij} = \gamma_0 + \gamma_1 x_{1ij} + \gamma_2 x_{2ij} + u_j + e_{ij},$$

where $u_j \sim N(0, \sigma_u^2)$ and $e_{ij} \sim N(0, \sigma_e^2)$ as usual. Here x_{1ij} and x_{2ij} are coded $+0.5$ and -0.5 for the condition of interest (e.g., intervention group) and the reference condition (e.g., control group). As a consequence, γ_1 is the difference in average outcome between the condition of interest and the reference condition for the first factor. This is irrespective of the condition of the second factor, provided interaction is absent. The same holds for γ_2 of course.

If a factor varies on cluster level, the variance of its effect estimator is

$$var(\hat{\gamma}) = 4\frac{n_1\sigma_u^2 + \sigma_e^2}{n_1 n_2} = 4\frac{[1 + (n_1 - 1)\rho]\sigma^2}{n_1 n_2},$$

and when a factor varies on subject level, the corresponding variance is

$$var(\hat{\gamma}) = 4\frac{\sigma^2}{n_1 n_2}$$

regardless of whether is it the first or the second factor. In the above, $\rho = \sigma_u^2/(\sigma_u^2 + \sigma_e^2)$ is the intraclass correlation coefficient and $\sigma^2 = \sigma_u^2 + \sigma_e^2$ is the total variance as usual.

As a consequence, sample size for a factor that varies on subject level can be calculated as in a multisite trial with a constant effect of treatment, and sample size for a factor that varies on cluster level can be calculated as in a cluster randomized trial.

10.5.2 Binary outcome

Again a restriction is made to models without interaction between the two factors; see Moerbeek and Maas (2005) for the case with interaction. In a factorial design with two factors, x_{1ij} and x_{2ij}, the probability p_{ij} of an dichotomous outcome $y_{ij} = 1$ for subject i in cluster j is modeled as

$$p_{ij} = \frac{1}{1 + \exp[-(\gamma_0 + \gamma_1 x_{1ij} + \gamma_2 x_{2ij} + u_j)]},$$

where the cluster level random term u_j has zero mean and variance σ_u^2. Again the factors x_{1ij} and x_{2ij} take the values -0.5 for their control and $+0.5$ for their intervention condition. The response $y_{ij} = p_{ij} + e_{ij}$ follows a Bernoulli distribution with expected value p_{ij} and hence the subject level random term e_{ij} has zero mean and variance $p_{ij}(1 - p_{ij})$.

For planning purposes, the variance component σ_u^2 may be assumed known and then γ_1 and γ_2 can be estimated by their generalized least squares estimators. Since the variances $var(e_{ij}) = p_{ij}(1 - p_{ij})$ in the four groups corresponding to the combinations of x_1 and x_2 are different, these estimators do not have a simple form and are hence omitted here. The variances of the estimators fortunately do take a relatively simply closed form. We introduce $\tilde{p}_{ij}$ to denote p_{ij} evaluated at $u_j = 0$, and $\sigma_{ij}^2 = \tilde{p}_{ij}(1 - \tilde{p}_{ij})$. Then we use notation σ_{++}^2, σ_{+-}^2, σ_{-+}^2, and σ_{--}^2 to denote the variances in the four combinations of the two factors. Here, the first and second subscript denote the levels of the first and second factor with notation "-" and "+" for the control and intervention levels.

If both factors are on cluster level, the variances of the effects of each factors are

$$\begin{aligned} var(\hat{\gamma}_1) &= 4\frac{[(\sigma_{--}^2 + \sigma_{+-}^2 + 2n_1\sigma_u^2)^{-1} + (\sigma_{-+}^2 + \sigma_{++}^2 + 2n_1\sigma_u^2)^{-1}]^{-1}}{n_1 n_2}, \\ var(\hat{\gamma}_2) &= 4\frac{[(\sigma_{--}^2 + \sigma_{-+}^2 + 2n_1\sigma_u^2)^{-1} + (\sigma_{+-}^2 + \sigma_{++}^2 + 2n_1\sigma_u^2)^{-1}]^{-1}}{n_1 n_2}. \end{aligned} \quad (10.41)$$

If the first factor is on cluster level and the second on subject level:

$$\begin{aligned} var(\hat{\gamma}_1) &= 4\frac{n_1\sigma_u^2+[(\sigma_{--}^2+\sigma_{+-}^2)^{-1}+(\sigma_{-+}^2+\sigma_{++}^2)^{-1}]^{-1}}{n_1 n_2}, \\ var(\hat{\gamma}_2) &= 4\frac{[(\sigma_{--}^2+\sigma_{-+}^2)^{-1}+(\sigma_{+-}^2+\sigma_{++}^2)^{-1}]^{-1}}{n_1 n_2}. \end{aligned} \tag{10.42}$$

Finally, if both factors are on individual level:

$$\begin{aligned} var(\hat{\gamma}_1) &= 4\frac{[(\sigma_{--}^2+\sigma_{+-}^2)^{-1}+(\sigma_{-+}^2+\sigma_{++}^2)^{-1}]^{-1}}{n_1 n_2}, \\ var(\hat{\gamma}_2) &= 4\frac{[(\sigma_{--}^2+\sigma_{-+}^2)^{-1}+(\sigma_{+-}^2+\sigma_{++}^2)^{-1}]^{-1}}{n_1 n_2}. \end{aligned} \tag{10.43}$$

See Moerbeek and Maas (2005) for a justification of these equations. In order to obtain closed formulas, the above variances for the effects of each factor were derived by a first order Taylor expansion around the fixed effects, resulting in so-called marginal quasi likelihood (MQL) estimators. As MQL estimators turn out to be biased downward, they are generally discouraged for fitting multilevel models to data. Estimators based on a second order Taylor expansion of the fixed and random effects, also called second order penalized quasi-likelihood (PQL) estimators, are less biased and therefore more often used in practice. From a simulation study (Moerbeek & Maas, 2005), the variance of second order PQL is on average 1.06 larger than the variance of MQL estimators for cluster level factors and 1.07 larger for subject level factors. Therefore, inflating the variance by MQL by a conversion factor 1.07 will on average give the variance for second order PQL estimation.

10.5.3 Sample size calculation for factorial designs

The following applies only to the case of a binary outcome with second order PQL estimation. However, sample size calculations for continuous outcomes are similar, provided that the subject level variance for continuous outcomes is used and of course the factor 1.07 to account for PQL estimation instead of MQL estimation is not needed.

Consider first the situation that both levels vary at cluster level. If the cluster size n_1 is fixed, the total number of clusters n_2 that is needed to detect an effect γ_1 for the first factor with power $1-\beta$ at two-sided significance level α can first be determined approximately using z-percentiles as

$$n_2 = 1.07\frac{(z_{1-\alpha/2}+z_{1-\beta})^2}{\gamma_1} \cdot \frac{[(\sigma_{--}^2+\sigma_{+-}^2+2n_1\sigma_u^2)^{-1}+(\sigma_{-+}^2+\sigma_{++}^2+2n_1\sigma_u^2)^{-1}]^{-1}}{n_1},$$

where the factor 1.07 arises to account for second order PQL estimation. Multiplying this initial n_2 by the small sample correction factor $(n_2+1)/(n_2-1)$ (approximately) accounts for the loss of power due to estimation of variance components (see p. 108 of Steel and Torrie (1980)). The resulting value should of course be rounded upward to the nearest integer multiple of 4 to get an integer number of clusters in each of the four combinations of the two binary factors.

Similarly, the total number of clusters required for detecting an effect γ_2 for the second (with power $1-\beta$ and two-sided significance level α) can be determined. The largest of the required total number of clusters for the first and second factor will provide sufficient power to detect effects in both factors.

In principle, when the total number of clusters n_2 is fixed, say $\bar{n}_2$, the minimal cluster size n_1 can be determined, but this will involve solving a quadratic equation in n_1, for example

$$1.07\frac{[(\sigma^2_{--}+\sigma^2_{+-}+2n_1\sigma^2_u)^{-1}+(\sigma^2_{-+}+\sigma^2_{++}+2n_1\sigma^2_u)^{-1}]^{-1}}{n_1\bar{n}_2} = \frac{\gamma_1^2}{(z_{1-\alpha/2}+z_{1-\beta})^2},$$

with no simple closed form. For practical purposes, it is therefore advised to list the required total number of clusters n_2 for various cluster sizes n_1 to find numerically the minimal cluster size n_1 that can be combined with $\bar{n}_2$ to obtain sufficient power to detect effects in both factors.

Sample sizes when at least one factor varies at subject level can be found similarly. Actually, for factors that vary at subject level, both the sample sizes n_2 and n_1 have the same expression with their roles exchanged, for example, for the second factor varying at subject level:

$$\begin{aligned} n_2 &= \frac{(z_{1-\alpha/2}+z_{1-\beta})^2}{\gamma_1} \cdot \frac{[(\sigma^2_{--}+\sigma^2_{-+})^{-1}+(\sigma^2_{+-}+\sigma^2_{++})^{-1}]^{-1}}{n_1}, \\ n_1 &= \frac{(z_{1-\alpha/2}+z_{1-\beta})^2}{\gamma_1} \cdot \frac{[(\sigma^2_{--}+\sigma^2_{-+})^{-1}+(\sigma^2_{+-}+\sigma^2_{++})^{-1}]^{-1}}{n_2}; \end{aligned}$$

see (10.42) and (10.43).

Other considerations. The above sample size calculations are based on either fixing n_1 or n_2 and then solving for the other. Another criterion to motivate a sample size could be to define the costs for sampling a cluster and for sampling a subject. Then, given a total budget, the allocation of clusters and subjects to yield the most precision and power can be determined. Or conversely, given a desired level of power or precision, the minimal total costs to achieve this can be calculated (Moerbeek & Maas, 2005; Moerbeek et al., 2003c).

11

The problem of unknown intraclass correlation coefficients

The previous chapters presented formulae for the purpose of calculating sample sizes to achieve a desired power level in a trial with nested data structures. In almost all cases the power was shown to depend, among other factors, on the size of the intraclass correlation coefficient. Unfortunately, the value of this model parameter is generally unknown in the design stage of a trial. This causes a vicious circle: a trial is designed and implemented to gain insight in the size of some model parameters, in particular the treatment effect, but to efficiently design a trial, the size of another model parameter, namely the intraclass correlation coefficient, needs to be known beforehand.

One way to escape this circle is to provide an a priori estimate of the intraclass correlation coefficient based on expert knowledge or previous research. The trouble with this approach is that there is no guarantee such a prior estimate is equal to the true value and as a result the trial may be under- or overpowered. This may be avoided by using an internal pilot study where an interim analysis is performed to re-estimate the intraclass correlation coefficient with the purpose of calculating the sample size in the remainder of the trial. Such studies are useful when the amount of time between the intervention and primary outcome is limited.

In studies where this is not the case, Bayesian sample size calculations may serve as an alternative. With such designs a prior distribution for the intraclass correlation coefficient is specified and a simulation study is performed to calculate posterior distributions for the required sample sizes. The sample size for the trial at hand can be chosen in such a way that the desired power level is achieved with sufficient probability.

Sample size calculations based on internal pilots and Bayesian techniques are useful when either the number of clusters or the cluster size is fixed and the other value needs to be calculated such that a pre-defined power level is achieved. If neither of these quantities is fixed beforehand, a cost function can be used to obtain the optimal sample sizes.

The optimal design depends on the value of the intraclass correlation coefficient and is called a locally optimal design. Using an incorrect estimate of the intraclass correlation coefficient results in a loss of efficiency and a maximin optimal design may be used to limit such a loss.

11.1 Estimates from previous research

Since the mid 1990s some 60 papers published estimates of intraclass correlation coefficients, mainly in the field of health promotion in the school setting or in primary or secondary care. An overview of these papers is given in Table 11.1. Although this overview is likely to be incomplete, it gives a list of relevant sources that may be consulted to gain insight into realistic values of the intraclass correlation coefficient for future trials. Researchers in any field, but especially the field of educational sciences, are encouraged to report intraclass correlation coefficient estimates when they report the outcomes of any trial with clustered data. This is also encouraged by item 17 of the extension of the CONSORT statement for cluster randomized trials (M. K. Campbell, Elbourne, & Altman, 2004).

M. K. Campbell, Grimshaw, and Elbourne (2004) describe the results of a survey to inform the appropriate reporting of the intraclass correlation coefficient. They identified three dimensions: a description of the data set, information on how the intraclass correlation coefficient was estimated and information on the precision of the estimate.

A description of the data set should consist of a demographic distribution between and within clusters, and a description of the outcome and intervention. With respect to the calculation of the ICC estimate, it is important to report which method has been used (for instance ANOVA or other method), the software used, whether the ICC was calculated from intervention and control data or from control data only and whether adjustments for covariates were made.

With respect to the precision of the ICC estimate it is important to report the confidence interval, the number of clusters, the average cluster size and the range of cluster sizes. Bonett (2002) provides sample sizes required to obtain confidence intervals of desired width; Bayesian methods to construct confidence intervals are presented by Turner, Omar, and Thompson (2001).

The width $2d$ of the $(1-\alpha)100\%$ confidence interval for ρ is approximately equal to $2z_{1-\alpha/2}\sqrt{var(\hat{\rho})}$, where for equal cluster sizes the approximate variance $var(\hat{\delta})$ is calculated as

$$var(\hat{\rho}) = \frac{2(1-\rho)^2(1+(n_1-1)\rho)^2}{(n_1-1)(n_1n_2-n_1)}. \tag{11.1}$$

This is a highly accurate approximation to the true variance of $\hat{\delta}$ when $n_2 \geq 30$ (Donner & Koval, 1983) and provides good coverage for all values $\rho \in [0, 0.9]$ (Donner & Wells, 1986). Given a fixed cluster size n_1 the number of clusters n_2 that yields a width $2d$ for the confidence interval is found by setting $2d = 2z_{1-\alpha/2}\sqrt{var(\hat{\rho})}$ and solving for n_2

$$n_2 = \frac{8z^2_{1-\alpha/2}((1-\rho)^2(1+(n_1-1)\rho)^2)}{n_1(n_1-1)4d^2} + 1. \tag{11.2}$$

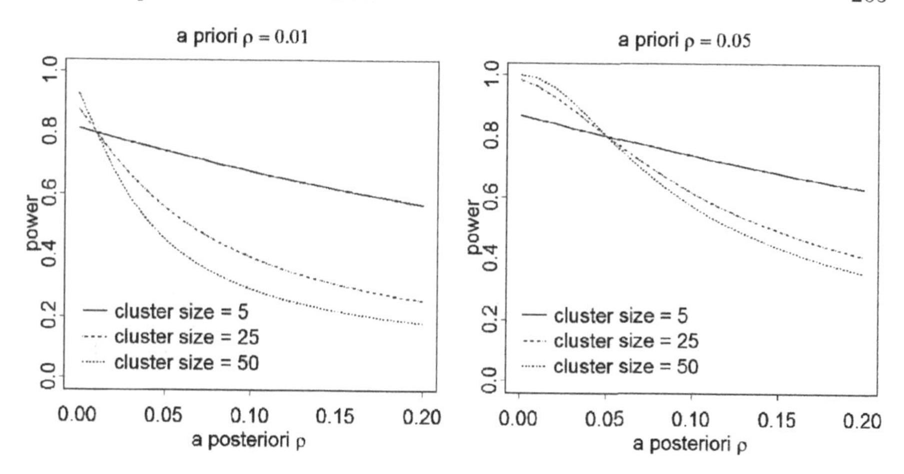

Figure 11.1: The effect of an incorrect a priori intraclass correlation coefficient estimate on power for the test on treatment effect.

In the planning phase of a trial the true but unknown value ρ should be replaced by a prior estimate.

Even when a very precise estimate of the intraclass correlation coefficient is used in the planning of a future trial, there is no guarantee this estimate equals true value for a future trial. Figure 11.1 shows the risk of an incorrect a priori estimate of the intraclass correlation coefficient on the power to detect a treatment effect in a cluster randomized trial with a fixed and common cluster size of 5, 25 or 50.

The number of clusters to achieve a power of 0.8 in a two-sided test with a significance level of 0.05 is calculated on the basis of Equation (4.13) for an a priori intraclass correlation of $\rho = 0.01$ or $\rho = 0.05$. The real power for such a design is then calculated for various values of the a posteriori estimated intraclass correlation coefficient within the range $[0, 0.20]$ and depicted in Figure 11.1. This figure clearly indicates that an underestimate of the intraclass correlation coefficient in the planning phase of a cluster randomized trial results in a underpowered study, especially when the cluster size is large (see also Guittet et al. (2005)). An overestimate results in an overpowered study. Only when the a priori intraclass correlation coefficient is equal to the a posteriori estimate is the desired power level of 0.8 achieved.

11.2 Sample size re-estimation

The sample size to achieve a desired power level almost always depends on model parameters that are unknown in the design phase of a trial but can

Table 11.1: Overview of literature that presents ICC estimates

Source	research field	subjects	clusters
Adams et al. (2004)	primary care	various	primary care practices and practitioners
Agarwal, Awasthi, and Walter (2005)	vitamin A intake	children	administrative blocks and villages
Amorim, Bangdiwala, McMurray, Creighton, and Harrell (2007)	physiology	children and adolescents	public schools
Baskerville, Hogg, and Lemelin (2000)	prevention in primary care	patients	primary care practices
Brandon, Harrison, and Lawton (2012)	education	students	schools
M. K. Campbell et al. (2001)	implementation	patients	primary and secondary care practices
M. K. Campbell, Fayers, and Grimshaw (2005)	implementation	patients	primary and secondary care practices
Carlin and Hocking (1999)	traffic, health	children and adolescents	schools
Elley, Kerse, Chondros, and Robinson (2005)	primary and residential health care	middle-aged and older adults	primary care practices and residential care homes
Feder, Griffiths, Eldridge, and Spence (1999)	secondary prevention of coronary heart disease	patients with acute coronary events	general practices
Feng et al. (1999)	various	various	various
Haddad et al. (2012)	perinatal care	pregnant women	hospitals
Gulliford, Ukoumunne, and Chinn (1999)	health care	adults	postal codes and households

Table 11.1 (continued)

Source	research field	subjects	clusters
Hannan, Murray, Jacobs, and McGovern Jr. (1994)	heart health	individuals	neighborhoods and cities
Hawkins, Van Horn, and Arthur (2004)	substance use	public school students	small to moderate sized towns
Hedberg and Hedges (2014)	education	pupils	schools and districts
Hedges and Hedberg (2007)	education	children, adolescents	schools
Hedges and Hedberg (2014)	education	students	schools
Hutchison (2009)	education	pupils	primary and secondary schools
Ip, Wasserman, and Barkin (2011)	pediatrics	parents	pediatric practices
Jacob, Zhu, and Bloom (2010)	education	students	classes and schools
Janega et al. (2004a)	nutrition intervention	students	schools
Janega et al. (2004b)	alcohol, tobacco, drugs	students	middle schools
Janjua, Khan, and Clemens (2006)	injection practices in developing countries	children, adults	groups within geographic boundaries
Kelcey and Phelps (2013)	education	teachers	schools
Kelcey and Phelps (2014)	education	teachers	schools
Knox and Chondros (2004)	primary care	patients	general practitioners
Konstantopoulos (2009)	education	students	classes and schools
Littenberg and MacLean (2006)	diabetes in primary care	adults with diabetes	primary care practices
Lajos et al. (2014)	perinatal care	births	referral hospitals
Martinson, Murray, Jeffery, and Hennrikus (1999)	worksite health promotion	employees	businesses

Table 11.1 (continued)

Source	research field	subjects	clusters
Metcalf, Scragg, Stewart, and Scott (2007)	nutrition	children	neighboring houses
Murray and Hannan (1990)	tobacco and drug use	students	junior and senior high schools
Murray et al. (1994)	adolescent smoking	students	schools
Murray and Short (1995)	alcohol use	young adults	communities
Murray and Short (1996)	alcohol use	pupils	schools
Murray and Short (1997)	tobacco use	adolescents	schools
Murray, Phillips, Birnbaum, and Lytle (2001)	nutrition	students	schools
Murray et al. (2002)	tobacco use	adolescents	classrooms
Murray, Van Horn, et al. (2006)	alcohol, tobacco and drug use	adolescents	communities
Murray, Stevens, et al. (2006)	physical activity	sixth grade girls	schools
Murray, Blitstein, Hannan, Baker, and Lytle (2007)	eating and nutrition	7th and 8th graders	middle schools
Pagel et al. (2011)	perinatal care	mothers and newborns	communities in developing countries
S. L. Pals, Beaty, Posner, and Bull (2009)	HIV/STD prevention	young African American and Hispanic females	neighborhoods
Parker, Evangelou, and Eaton (2005)	cholesterol education and research	patients	primary care practices
Piaggio et al. (2001)	antenatal care	pregnant women	clinics
Preisser, Reboussin, Song, and Wolfson (2007)	underage drinking	adolescents	cities and counties
Reading, Harvey, and McLean (2000)	maternal and child health	families with infants under 1 year	practices

Table 11.1 (continued)

Source	research field	subjects	clusters
Roudsari, Nathens, Koepsell, Mock, and Rivara (2006)	trauma	patients	trauma centers
Roudsari, Fowler, and Nathens (2007)	childhood trauma	children	trauma centers
Resnocow et al. (2010)	smoking	pupils	high schools
Rowe, Lama, Onikpo, and Deming (2002)	health care	patients	health facility
Scheier, Griffin, Doyle, and Botvin (2002)	drug abuse	students	schools
Schochet (2008)	education	students	schools
Siddiqui et al. (1996)	smoking	students	classes and schools
Slymen and Hovell (1997)	tobacco and alcohol use	adolescents	orthodontists
Slymen et al. (2003)	tobacco and alcohol use	migrant adolescents	schools
Smeeth and Ng (2002)	assessment and management of older people	adults 75 years and older	general practices
Taljaard et al. (2008)	maternal and perinatal health	mothers and newborns	hospitals
D. M. Thompson, Fernald, and Mold (2012)	health care	patients	general practices
Westine, Spybrook, and Taylor (2014)	education	students	schools
Xu and Nichols (2010)	education	students	classes and schools
Yelland, Salter, Ryan, and Laurence (2011)	pathology testing	patients	primary care practices
Zhu, Jacob, Bloom, and Xu (2012)	education	students	classes and schools

be estimated during data collection. The data used to calculate the model parameters are called pilot data. Stein (1945) was the first to propose sample size re-estimation. In his procedure the data to estimate the model parameters (in his case variances) were not used in the final analysis. Wittes and Brittain (1990) proposed an adjustment that uses all data in the final analysis, including the data used to estimate the model parameters. These pilot data are now called internal pilot data and the advantage of this approach is that as all data are used in the final analysis, a higher level of power can be achieved. The other side of the coin is that the type I error rate may not be preserved because of the dependence between the estimates from the internal pilot and the final sample size (Wittes & Brittain, 1990; Wittes, Schabenberger, Zucker, Brittain, & Proschan, 1999). However, recent research indicates that under general conditions the type I error rate is preserved (Broberg, 2013). For a review the reader is referred to Friede and Kieser (2006) and Proschan (2009). Sample size re-estimation can be regarded a special case of adaptive designs; a special issue on this type of designs appeared in *Biometrical Journal* (Röhmel, 2006).

Sample size re-estimation has been studied for multicenter trials (Jensen & Kieser, 2010) and cluster randomized trials (Lake et al., 2002). In the study by Jensen and Kieser (2010), the number of centers was four and the number of patients per center was calculated on basis of the pilot data in these centers. In the study by Lake et al. (2002), data from the clusters in the pilot were used to calculate how many additional clusters should be recruited; no additional subjects from the clusters in the internal pilot were sampled. The steps of an internal pilot design for the latter study are summarized as follows:

1. Specify the alternative hypothesis H_a, effect size γ_1, desired power level $1-\beta$ and type I error rate α.
2. Decide on the number of subjects to be sampled per cluster.
3. Give prior estimates of the intraclass correlation coefficient based on expert knowledge or findings from the literature.
4. Calculate the number of clusters n_2 by means of Equation (4.9).
5. Sample a proportion π of these clusters, implement the intervention and measure the outcome variable.
6. Re-estimate the intraclass correlation coefficient on basis of the pilot data and re-calculate the required number of clusters.
7. If the number of clusters as calculated in the previous step is larger than the number of clusters in the pilot, sample additional clusters, implement the intervention and measure the outcome.
8. Perform the test on treatment effect using all data including the pilot data.

The above assumes equal numbers of subjects are sampled from each cluster. When this is not the case it suffices to sample 10% more clusters to compensate for the loss in efficiency (Van Breukelen et al., 2007). The steps above can be

Table 11.2: Empirical type I error rate α and power $1-\beta$ for standard design and re-estimation design.

	standard design		re-estimation design					
			$\pi = 0.25$		$\pi = 0.5$		$\pi = 0.75$	
prior ρ	α	$1-\beta$	α	$1-\beta$	α	$1-\beta$	α	$1-\beta$
0.025	0.0516	0.6896	0.0624	0.7834	0.0588	0.7994	0.0556	0.8020
0.050	0.0504	0.7996	0.0568	0.8000	0.0506	0.8052	0.0518	0.8166
0.075	0.0482	0.8810	0.0584	0.8006	0.0566	0.8122	0.0518	0.8258

adjusted if the number of clusters is fixed beforehand and a decision needs to be made on the number of subjects per cluster (Van Schie & Moerbeek, 2014). In that case, the number of clusters may be too small to achieve the desired power level based on the estimate of the intraclass correlation coefficient from the internal pilot (see also Section 4.3.1). One should then decide to include additional clusters or accept a lower power level.

Table 11.2 compares the type I error rates and power levels of a trial with sample size re-estimation to those without (i.e., the standard design) as found in a simulation study with 5000 generated data sets for each condition. A small effect size ($\delta = 0.2$) was used and the true value of the intraclass correlation coefficient used to generate the data was $\rho = 0.05$. Three prior estimates of the intraclass correlation coefficient and three values of the proportion π were used. Sample size calculations and statistical tests were performed at $\alpha = 0.05$ and the cluster size was fixed to $n_1 = 25$.

The table shows that while the type I error rates for the standard design are close to their nominal value, those for the internal pilot design are slightly inflated. The degree of inflation is largest when $\pi = 0.25$. The power levels for the standard design are only close to the desired level when the prior estimate of the intraclass correlation coefficient is equal to the true value. Too small prior estimates result in an underpowered study while an overpowered study is the result of a prior estimate that is too high. The power is very well preserved for the internal pilot design. For prior $\rho = 0.025$ and $\pi = 0.25$ the size of the internal pilot is too small to result in a good estimate of the true ρ and this may result in a study that is slightly underpowered. For prior $\rho = 0.025$ and $\pi = 0.75$ the size of the internal pilot is already too large, which results in a study that is slightly overpowered.

11.3 Bayesian sample size calculation

The main advantage of the Bayesian approach to sample size calculation is that it can implicitly take uncertainty about the intraclass correlation coefficient into account by using a prior distribution of this model parameter. A large number of draws of the intraclass correlation coefficient are generated from this prior distribution and for each value the required sample sizes are calculated such that a desired power level is achieved. One may then select an upper percentile from the distribution of sample sizes to be reasonably confident that the study has adequate power.

Conversely, one may start with a given sample size at the cluster and subject levels and for each of the draws from the prior distribution of the intraclass correlation coefficient calculate the power level that can be achieved. The prior distribution may be based on subjective beliefs (Spiegelhalter, 2001) or a single or collection of estimates (Rotondi & Donner, 2009; Turner, Prevost, & Thompson, 2004; Turner, Thompson, & Spiegelhalter, 2005). With the latter, the relevance and precision of each estimate can be taken into account.

It should be mentioned that a Bayesian perspective is adopted to specify beliefs about the plausible values of the intraclass correlation coefficient but it is assumed that analysis of the data will be done in the frequentist framework. For this reason, this approach to power calculation has been labeled hybrid Neyman-Pearson and Bayesian (Spiegelhalter, Abrams, & Myles, 2004).

Consider as an example a cluster randomized trial with a common cluster size of $n_1 = 30$. Let us assume the intraclass correlation to be 0.025. To achieve 80% power to detect a small standardized treatment effect $\delta = 0.2$ in a two-sided test with type I error rate $\alpha = 0.05$ a total of 48 clusters is needed. Uncertainty about the value of the intraclass correlation coefficient can be specified by a prior distribution. Let us assume this parameter is about 0.025 but there is some chance it is as large as 0.10. This uncertainty may be reflected by a normal distribution with a mean 0.025 and standard deviation 0.01, but truncated at zero so that negative values are excluded. The corresponding prior density is shown in the top panel of Figure 11.2. A sample of 100,000 values is drawn from this distribution and for each the required number of clusters is calculated on basis of (4.13) and (4.9). The corresponding density is shown in the middle panel of Figure 11.2 and it is observed that in many cases a sample size at the cluster level larger than $n_2 = 48$ would be required. Conversely, a design with 48 clusters may not always result in the desired power level, as is shown in the bottom panel of Figure 11.2. On basis of these findings one may decide to use more than 48 clusters to be reasonably confident the study has sufficient power. For instance, the 95th percentile of the prior distribution of n_2 may be selected, which in this example is equal to $n_2 = 60$ clusters.

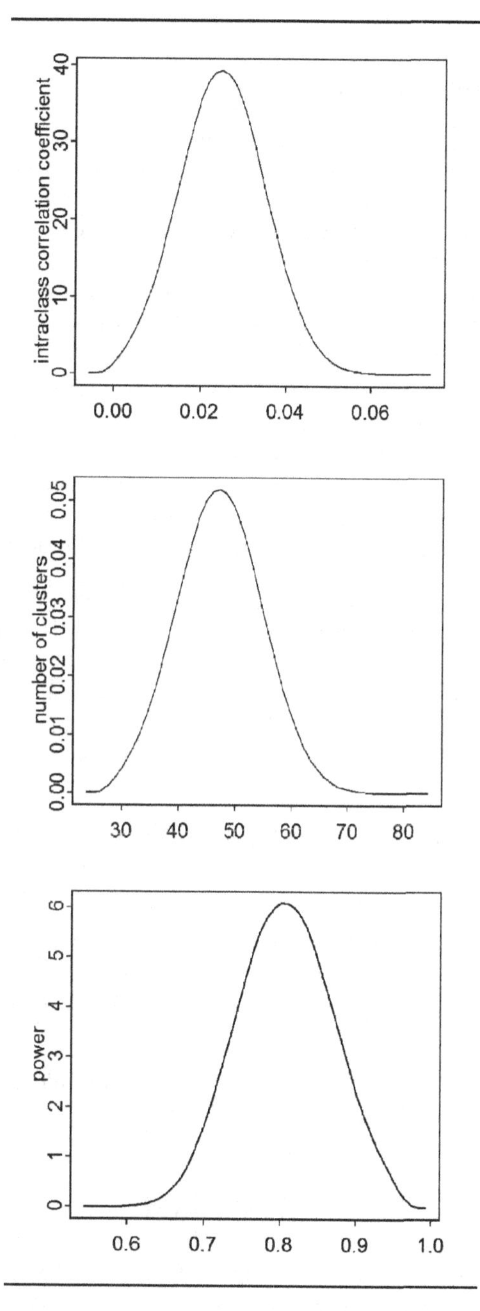

Figure 11.2: Densities of prior distribution of intraclass correlation coefficient, required number of clusters to receive a power level 0.8, and power at 48 clusters.

11.4 Maximin optimal designs

Bayesian optimal designs and designs that use internal pilots are useful in case the number of clusters or cluster size is fixed beforehand and the other value needs to be calculated such that a desired power level is achieved. In case neither sample size is fixed beforehand a cost function can be used to calculate the optimal design; see Section 4.3.4. As the optimal sample sizes depend on the intraclass correlation coefficient, the design is called a locally optimal design. This implies a design that is optimal for some intraclass correlation coefficient is not optimal for other values and an incorrect estimate of the intraclass coefficient results in a loss of efficiency. This loss is still acceptable if one is able to provide an estimate that does not underestimate the true value by more than 75% or overestimate it by more than 175% (Korendijk et al., 2010). If one may be unable to provide an estimate within these limits then a maximin optimal design might be used to limit the loss in efficiency.

To find the maximin optimal design, three steps should be taken

1. Define the parameter and design space. The parameter space is the range of plausible values of the intraclass correlation coefficient ρ and denoted $[\rho_L, \rho_U]$, where ρ_L and ρ_U are the lower and upper limits of this range. The design space is the collection of designs $\xi = (n_1, n_2)$ that are feasible for the trial at hand. For instance, a budgetary constraint may be used to define the feasible design region.
2. For each $\rho \in [\rho_L, \rho_U]$ calculate the locally optimal design ξ^*. Next, compute the relative efficiency *RE* of each design ξ in the design space compared to the locally optimal design ξ^*.
3. For each design ξ in the design space select its minimum relative efficiency *RE*. Then select the design that maximizes the smallest *RE*. This is the maximin optimal design *MMD* and the minimum *RE* is called the maximin value *MMV*.

The parameter space $[\rho_L, \rho_U]$ is a continuum and for calculation of the maximin optimal design a grid should be used with a step size that is sufficiently small. The sample sizes n_1 and n_2, on the other hand, can only take integer values.

Van Breukelen and Candel (in press) derived the maximin design for trials where the design space is determined by the budgetary constraint $c_1 n_1 n_2 + c_2 n_2 \leq C$, where c_1 and c_2 are the costs at the subject and cluster level and C is the total budget; see also Section 4.3.4. They found the *MMD* has the following cluster size

$$n_1 = \frac{(\rho_U - 1)g(\rho_L) - (\rho_L - 1)g(\rho_U)}{\rho_U g(\rho_L) - \rho_L g(\rho_U)}, \tag{11.3}$$

with

$$g(\rho) = (\sqrt{\rho_L c_2} + \sqrt{(1-\rho_L)c_1})^2, \tag{11.4}$$

and the number of clusters is calculated as $n_2 = C/(c_1 n_2 + c_2)$. The cluster size of the maximin optimal design depends on the boundaries of the parameter space $[\rho_L, \rho_U]$ and the cost ratio c_2/c_1, but not on the costs c_1 and c_2 and the total budget C. This can be verified by dividing $g(\rho_L)$ and $g(\rho_U)$ by the subject level costs c_1. The minimum relative efficiency of the *MMD* is attained at either of the boundaries $\rho = \rho_L$ and $\rho = \rho_U$ and this value is equal to

$$MMV = \frac{n_1 g(\rho_L)}{(n_1\rho_L + (1-\rho_L))(c_2 + c_1 n_1))}. \tag{11.5}$$

Figure 11.3 shows how the maximin optimal design is derived graphically. The same costs at the subject and cluster level as in Section 4.3.4 are used: $c_1 = 10$ and $c_2 = 262$. The design space restricts to values n_1 between 4 and 60.

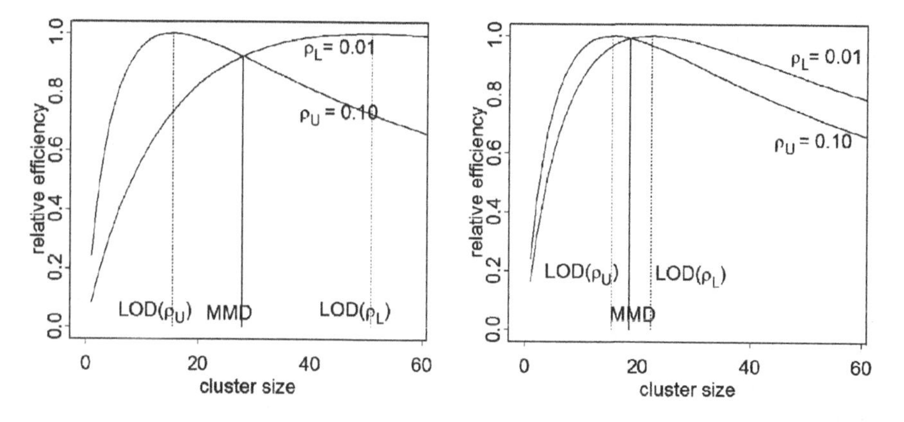

Figure 11.3: Graphical derivation of maximin optimal designs.

In the left graph, the parameter space is $[\rho_L, \rho_U] = [0.01, 0.1]$ and the two curves represent the relative efficiencies of the upper and lower boundary of this range across the parameter space. The relative efficiencies for all other values ρ in the design space are found between these two curves and are not shown. The locally optimal design for ρ_L is found at $n_1 = 50.9$ and that for ρ_U at $n_1 = 15.4$ and these are indicated by dashed vertical lines. The maximin optimal design is found at the intersection of the two curves: $n_1 = 27.81$ with a corresponding $MMV = 0.92$ and it is indicated by a solid vertical line.

Note that the intersection is not necessarily found at the mean value of the locally optimal designs for ρ_L and ρ_U. For the right graph, a larger lower bound $\rho_L = 0.05$ is used, which results in a narrower parameter space. In other words, the degree of uncertainty about the true value ρ is lower in the right graph. The locally optimal design for the lower bound ρ_L is found at

$n_1 = 22.3$ and that for the upper bound ρ_U remains unchanged at $n_1 = 15.4$. The maximin optimal design is found at $n_1 = 18.6$ and has an $MMV = 0.99$. The two locally optimal designs in the right graph are nearer than those in the left and as a result the maximin optimal design has a higher MMV.

This section has only given a short introduction to maximin optimal designs. For a more extensive overview including multicenter trials and trials with dichotomous endpoints the reader is referred to Van Breukelen and Candel (in press).

12

Computer software for power calculations

12.1 Introduction

The previous chapters presented mathematical equations to calculate sample sizes to achieve a desired power level in a study with hierarchical data. Sample sizes were derived to calculate the required number of clusters given a fixed cluster size or calculate the required cluster size given a fixed number of clusters. Attention was also paid to designs where neither sample size was fixed beforehand and a cost function was used to find the least expensive design to achieve a desired power level, or to derive the design with the highest power level on a fixed budget.

The sample sizes shown in this book are sometimes fairly simple, such as Equation (4.13), and sometimes far more complicated, such as Equation (8.12). To facilitate the work of statisticians and substantive researchers, it is important that user friendly software is available to perform power calculations in trials with hierarchical data structures.

Some computer programs are already available, such as PinT (Bosker, Snijders, & Guldemond, 2003). This is a general program that can be used for any study with two-level data, and is not restricted to experimental designs. This program calculates standard errors of the regression coefficients based on user-specified values for the means, variances and covariances of the predictor variables and variances and covariances of the residuals. The user is requested to specify the budget and the ratio of the costs at the cluster and subject level. Standard errors for the regression coefficients are calculated for a variety of combinations that satisfy the budgetary constraint (4.16) and the user can select the design that has smallest standard error for the regression coefficient of primary interest. The non-centrality parameter for the test of this regression coefficient is then calculated by dividing the value of the regression coefficient by its standard error and the power follows from Equations (3.6) through (3.8). As the true value of the regression coefficient is generally unknown, it has to be replaced by the minimally relevant effect size.

The Optimal Design software (Spybrook et al., 2011) mainly applies to experimental designs with two or three levels of nesting, longitudinal studies and meta analysis. It produces plots for the relation between power and a user-selected design factor as illustrated in Section 4.3.1 for cluster randomized trials. For cluster randomized and multisite trials the software also allows for the use of a budgetary constraint to calculate the optimal design.

The similarity between the PinT and Optimal Design programs is that they calculate the standard errors or power levels on the bases of mathematical equations.

Other programs rely on simulation studies and are described in more detail in Section 3.4. MLPowSim (W. J. Browne, Golalizadeh Lahi, & Parker, 2009) allows the user to specify a statistical model and generates macros and functions that can be used for simulation studies in the packages MLwiN or R. The simulation tool ML-DEs (Cools, 2008) can be used in a similar way: it generates macros to perform simulation studies in MLwiN and R scripts to handle the output of the simulations in R. The Mplus program was originally designed for data analysis but it also has extensive facilities to perform simulation studies for multilevel data (L. K. Muthén & Muthén, 2002).

In the remainder of this chapter a new program to perform power calculations and derive optimal designs is presented. This program allows calculation of required sample sizes for most designs in this book. The current version of the program is restricted to continuous outcome variables.

12.2 Computer program SPA-ML

The SPA-ML (Statistical Power Analysis for Multi-Level designs) program was written in MATLAB® but can be used as a Windows stand-alone program once the MATLAB Compiler Runtime (MCR) library has been installed on the computer. This file is downloaded from the Internet during the installation process. The installation package SPAMLInstaller_web.exe should be downloaded from http://tinyurl.com/SPAML. After double clicking on this file, all the files to support running the SPA-ML program are extracted and the program is added to the start menu. Upon request a shortcut can be added to the desktop. After the installation has been completed, the program can be started by double clicking SPA_ML.exe.

The first screen that appears is the main menu in Figure 12.1. The design for which a power calculation needs to be performed is chosen in step 1 and refined in step 2. The choices available in step 2 depend on the design chosen in step 1. The sample size scenario is selected in step 3 and the choices depend on the choices made in step 1 and 2. In Figure 12.1 a choice is made to calculate the number of clusters for a fixed cluster size in a cluster randomized trial with two levels of nesting. Tables 12.1 to 12.6 give an overview of all sample size scenarios in the SPA-ML program. The last columns of these tables list the equations in this book that are used in the SPA-ML program. Most of these equations are based on the normal approximation of the t distribution, but SPA-ML refines the calculated sample size by actually using the t distribution. As a consequence, the results obtained from SPA-ML may be slightly different from those based on the equations in this book, especially when the number of clusters is small. The only exception is made for three-level cluster randomized trials and multisite cluster randomized trials where a cost function is used to calculate the number of sites, clusters within sites and subjects within clusters. Finding the optimal design based on the t distribution is too computer-intensive in these cases and the normal approximation is used instead.

After pressing the calculate button at the bottom of the main menu a second screen appears; see Figure 12.2. This screen contains two tabs: the power protocol tab and the power graph tab. At the top of the left side of the power protocol tab the user needs to specify the input test parameters: whether a one-sided or two-sided alternative hypothesis is used, the type I error rate, the desired power level, the intraclass correlation coefficient and the cluster size. The notation corresponds to the notation that is used in this book.

Halfway down the left side column the effect size needs to be specified. One can either specify the standardized effect size or let the program calculate it from the mean and common standard deviation. The output appears once the calculate button is pressed. The output is presented in three formats: in output boxes at the bottom left, as a summary of power analysis in the text box at the top right and as a power statement in the text box at the bottom right. The power statement is very useful for project proposals and grant applications.

The text in the text boxes can be selected and copied to a text editor using the control+a and control+c key combinations. The second tab shows a power graph; see Figure 12.3. Here one can select three values of the standardized effect size, cluster size and intraclass correlation. This is very useful in case one is not certain about the true values of the standardized effect size and intraclass correlation and wants to evaluate and compare the power levels achieved with various plausible values of these parameters. The power graph that appears after pressing the button can be opened in a separate file by pressing the right mouse button and can then be printed or saved for further use.

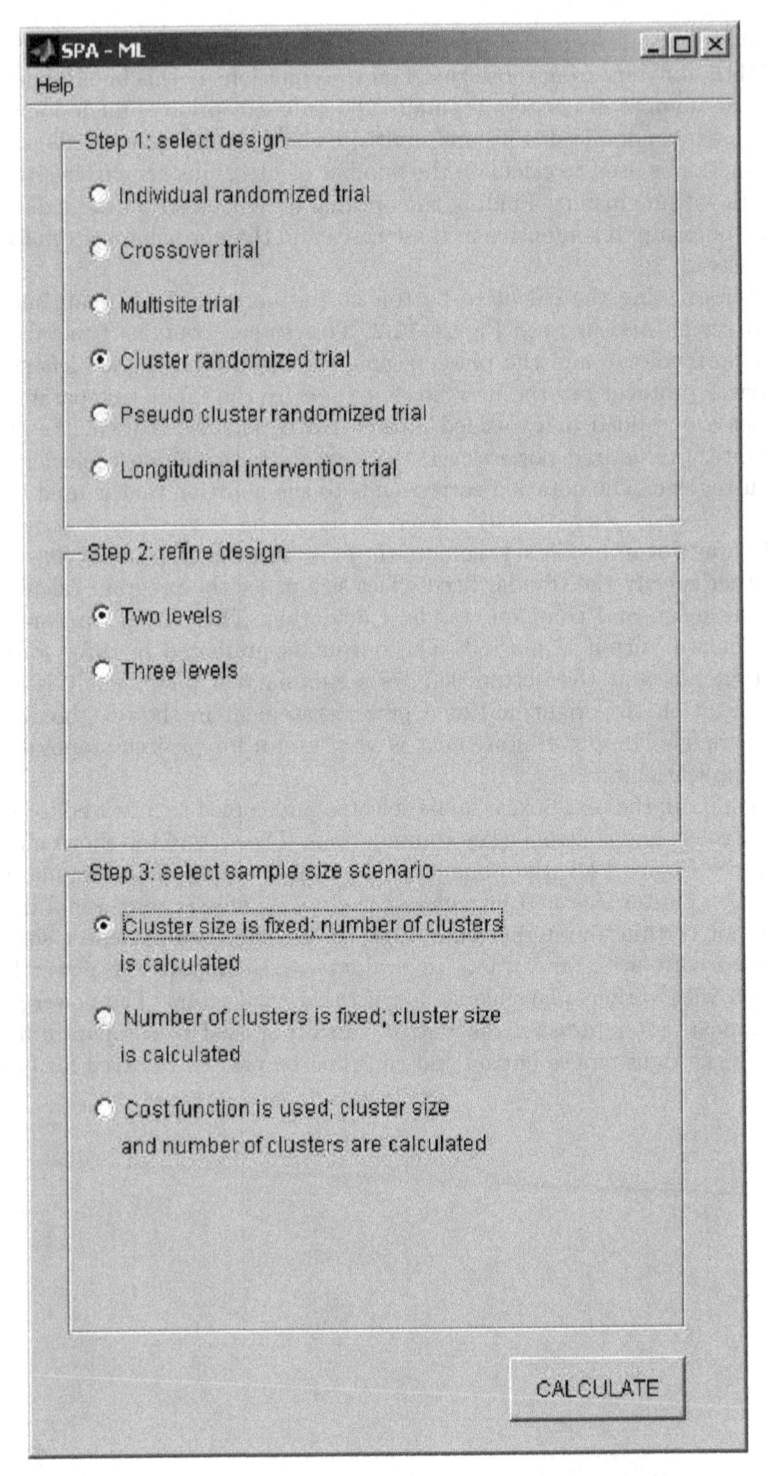

Figure 12.1: Main menu window of SPA-ML program.

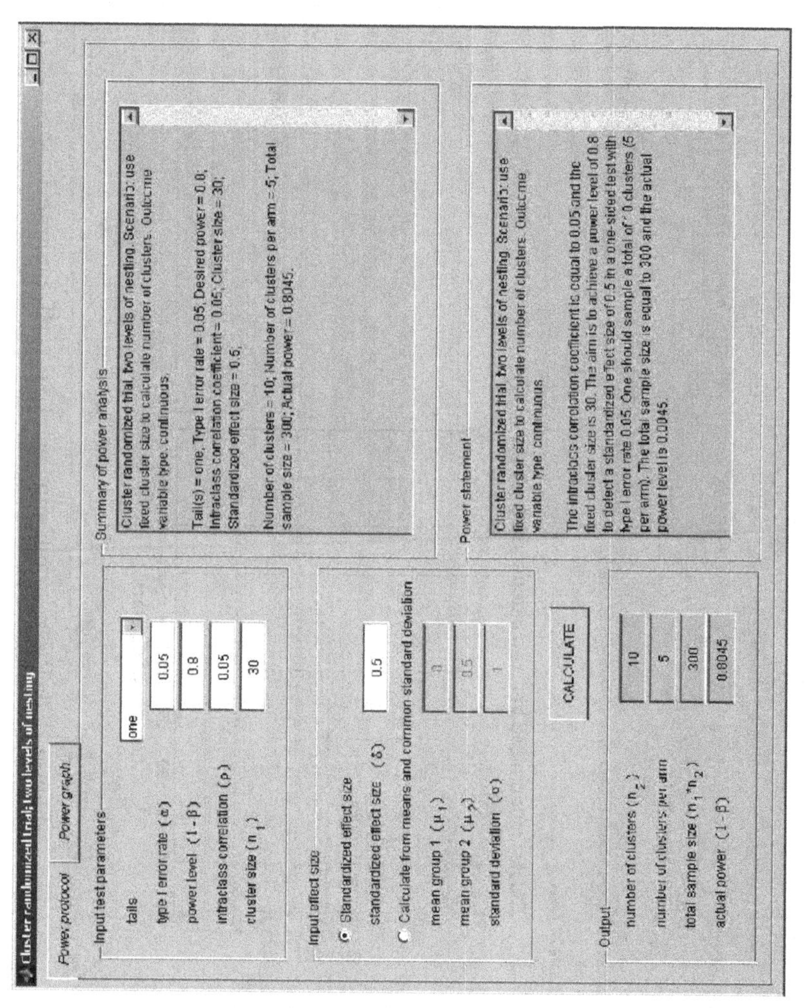

Figure 12.2: Power protocol window of SPA-ML program.

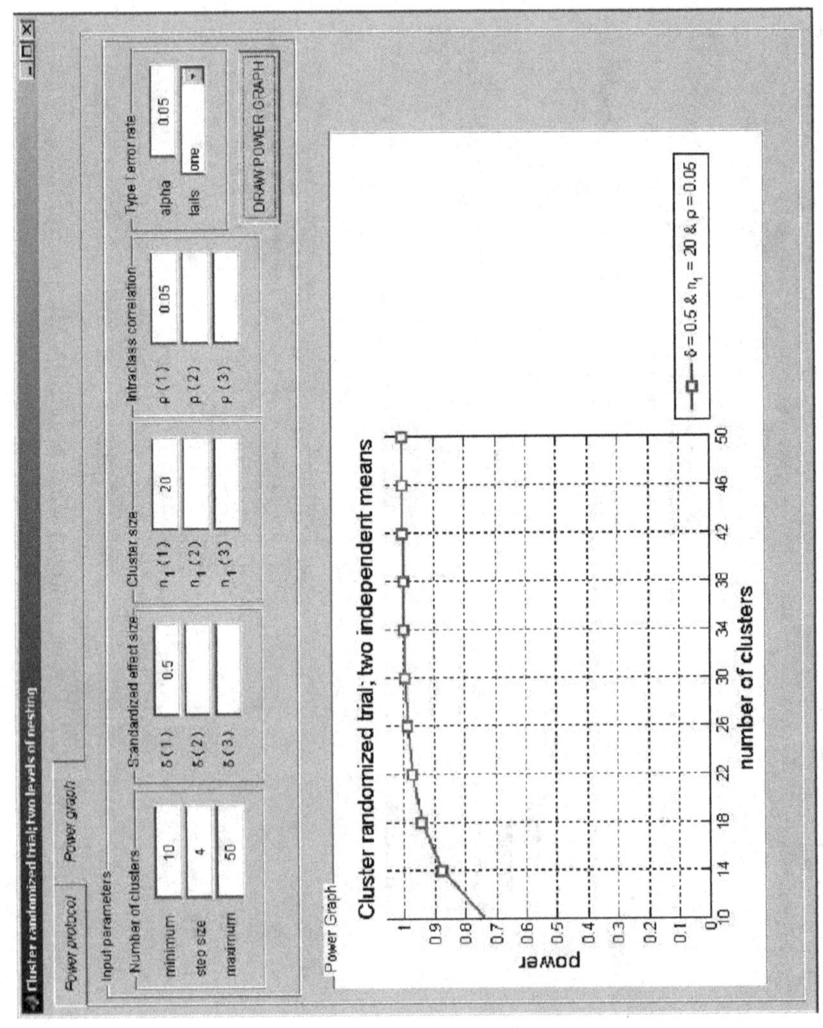

Figure 12.3: Power graph window of SPA-ML program.

Table 12.1: Overview of designs for individually randomized trials available in SPA-ML program

clustering in	scenario	equations
neither condition	(a) sample size in one group is fixed; sample size in other group is calculated, (b) sample size ratio is fixed; sample sizes in both groups are calculated, (c) cost function is used; sample sizes in both groups are calculated	(3.4)
one condition	sample size in control group and experimental cluster size are fixed; number of experimental clusters is calculated	(8.15)
	cluster size and number of clusters in experimental group are fixed; sample size in control group is calculated	(8.14)
	cost function is used; sample sizes in both groups are calculated	(8.18)
both conditions (i.e., group treatment)	number of experimental clusters is fixed; number of control clusters is calculated	(8.9)
	cost function is used; numbers of clusters in both conditions are calculated	(8.12)

Table 12.2: Overview of designs for crossover trials available in SPA-ML program

design	scenario	equations
individually randomized	(a) sample size in sequence AB is fixed; sample size in sequence BA is calculated,	(5.11)
	(b) sample size ratio is fixed; sample sizes in both sequences are calculated,	
	(c) cost function is used; sample sizes in both sequences are calculated	
cluster randomized cluster crossover	cluster size is fixed; number of clusters is calculated	(5.18)
	number of clusters is fixed; cluster size is calculated	(5.19)
	cost function is used; cluster size and number of clusters are calculated	(5.20), (5.21)
cluster randomized individual crossover	cluster size is fixed; number of clusters is calculated	(5.25)
	number of clusters is fixed; cluster size is calculated	(5.26)
	cost function is used; cluster size and number of clusters are calculated	–
stepped wedge	cluster size is fixed; number of clusters is calculated	–
	number of clusters is fixed; cluster size is calculated	–

Table 12.3: Overview of designs for multisite trials available in SPA-ML program

refinement	scenario	equations
two levels	cluster size is fixed; number of clusters is calculated	(6.16)
	number of clusters is fixed; cluster size is calculated	(6.17)
	cost function is used; cluster size and number of clusters are calculated	(6.21), (6.22)
three levels	(a) cluster size and number of clusters are fixed; number of sites is calculated, (b) cluster size and number of sites are fixed; number of clusters is calculated, (c) number of clusters and number of sites are fixed; cluster size is calculated	(10.14), Table 10.2
	cost function is used; cluster size, number of clusters and number of sites are calculated	(10.5), (10.15)

Table 12.4: Overview of designs for cluster randomized trials available in SPA-ML program

refinement	scenario	equations
two levels	cluster size is fixed; number of clusters is calculated	(4.13)
	number of clusters is fixed; cluster size is calculated	(4.14)
	cost function is used; cluster size and number of clusters are calculated	(4.18), (4.18)
three levels	(a) cluster size and number of clusters are fixed; number of sites is calculated, (b) cluster size and number of sites are fixed; number of clusters is calculated, (c) number of clusters and number of sites are fixed; cluster size is calculated	(10.3), Table 10.1
	cost function is used; cluster size, number of clusters and number of sites are calculated	(10.5), (10.6), (10.7)

Table 12.5: Overview of designs for pseudo cluster trials are available in SPA-ML program

refinement	scenario	equation
two levels of nesting	cluster size is fixed; number of clusters is calculated	(7.5)
	number of clusters is fixed; cluster size is calculated	(7.5)

Table 12.6: Overview of designs for longitudinal trials available in SPA-ML program

refinement	scenario	equation
linear relation between time and response	duration of the study and number of measurements are fixed; number of subjects is calculated	(9.11)
	(a) duration of the study and number of subjects are fixed; number of measurements is calculated, (b) number of subjects and number of measurements are fixed; duration of the study is calculated, (c) cost function is used; number of subjects, duration and number of measurements are calculated	(9.7)
quadratic relation between time and response	duration of the study and number of measurements are fixed; number of subjects is calculated	(9.11)
	(a) duration of the study and number of subjects are fixed; number of measurements is calculated, (b) number of subjects and number of measurements are fixed; duration of the study is calculated, (c) cost function is used; number of subjects, duration and number of measurements are calculated	(9.7)
cubic relation between time and response	duration of the study and number of measurements are fixed; number of subjects is calculated	(9.11)
	(a) duration of the study and number of subjects are fixed; number of measurements is calculated, (b) number of subjects and number of measurements are fixed; duration of the study is calculated, (c) cost function is used; number of subjects, duration and number of measurements are calculated	(9.7)

References

Achenbach, T. (1991). *Manual for the child behavior checklist 4-18, 91 profile.* Burlington, VT: University of Vermont, Department of Psychiatry.

Adams, G., Gulliford, M. C., Ukoumunne, O. C., Eldridge, S., Chinn, S., & Campbell, M. J. (2004). Patterns of intra-cluster correlation from primary care research to inform study design and analysis. *Journal of Clinical Epidemiology*, *57*(8), 785-794.

Afshartous, D. (1995). *Determination of sample size for multilevel model design.* (Unpublished manuscript)

Agarwal, G. G., Awasthi, S., & Walter, S. D. (2005). Intra-class correlation estimates for assessment of vitamin A intake in children. *Journal of Health, Population and Nutrition*, *23*(1), 66-73.

Ahn, C., Hu, F., Skinner, C. S., & Ahn, D. (2009). Effect of imbalance and intracluster correlation coefficient in cluster randomization trials with binary outcomes when the available number of clusters is fixed in advance. *Contemporary Clinical Trials*, *30*(4), 317-320.

Ahn, C., Hu, F. B., & Skinner, C. (2009). Effect of imbalance and intracluster correlation coefficient in cluster randomized trials with binary outcomes. *Computational Statistics and Data Analysis*, *53*(3), 596-602.

Allison, D. B. (1995). When is it worth measuring a covariate in a randomized clinical trial? *Journal of Consulting and Clinical Psychology*, *63*(3), 339-343.

Allison, D. B., Allison, R. L., Faith, M. S., Paultre, F., & Pi-Sunyer, F. X. (1997). Power and money: designing statistically powerful studies while minimizing financial costs. *Psychological Methods*, *2*(1), 20-33.

Amorim, L. D., Bangdiwala, S. I., McMurray, R. G., Creighton, D., & Harrell, J. (2007). Intraclass correlations among physiologic measures in children and adolescents. *Nursing Research*, *56*(5), 355-360.

Arnold, B. F., Hogan, D. R., Colford, J. M., & Hubbard, A. E. (2011). Simulation methods to estimate design power: an overview for applied research. *BMC Medical Research Methodology*, *11*, 94.

Atenafu, E. G., Hamid, J. S., Stephens, D., To, T., & Beyene, J. (2009). A small *p*-value from an observed data is not evidence of adequate power for future similar-sized studies: A cautionary note. *Contemporary Clinical Trials*, *30*(2), 155-157.

Atkinson, A. C., Donev, A. N., & Tobias, R. (2007). *Optimum experimental design, with SAS.* Oxford: Clarendon.

Ausems, M., Mesters, I., Van Breukelen, G., & De Vries, H. (2002). Short-term effects of a randomized computer-based out-of-school smoking prevention trial aimed at elementary school children. *Preventive Medicine*, *34*(6), 581-589.

Aydin, B., Leite, W. L., & Algina, J. (2014). The consequences of ignoring variability in measurement occasions within data collection waves in latent growth models. *Multivariate Behavioral Research*, *49*(2), 149-160.

Baldwin, S. A., Bauer, D. J., Stice, E., & Rohde, P. (2013). Evaluating models for partially clustered designs. *Psychological Methods*, *16*(2), 149-165.

Baskerville, N., Hogg, W., & Lemelin, J. (2000). The effect of cluster randomization in sample size in prevention research. *The Journal of Family Practice*, *50*(3), 241-246.

Bauer, D. J., Sterba, S. K., & Hallfors, D. D. (2008). Evaluating group-based interventions when control participants are ungrouped. *Multivariate Behavioral Research*, *43*(2), 210-236.

Bausell, R. B., & Li, Y. (2002). *Power analysis for experimental research: A practical guide for biological, medical and social sciences.* Cambridge: Cambridge University Press.

Berger, M. P. F., & Wong, W. K. (2009). *An introduction to optimal designs for social and biomedical research.* Chichester: Wiley.

Berkhof, J., & Kampen, J. K. (2004). Asymptotic effect of misspecification in the random part of the multilevel model. *Journal of Educational and Behavioral Statistics*, *29*(2), 201-218.

Berkhof, J., & Snijders, T. A. B. (2001). Variance component testing in multilevel models. *Journal of Educational and Behavioral Statistics*, *26*(2), 133-152.

Biau, D. J., Halm, J. A., Ahmadieh, H., Capello, W. N., Jeekel, J., Boutron, I., & Porcher, R. (2008). Provider and center effect in multicenter randomized controlled trials of surgical specialties: an analysis on patient-level data. *Annals of Surgery*, *247*(5), 892-898.

Biau, D. J., Porcher, R., & Boutron, I. (2008). The account for provider and center effects in multicenter interventional and surgical randomized controlled trials is in need of improvement: a review. *Journal of Clinical Epidemiology*, *61*(5), 435-439.

Bland, J. M. (2004). Cluster randomised trials in the medical literature: two bibliometric surveys. *BMC Medical Research Methodology*, *4*, 21.

Bloom, H. S. (2005). Learning more from social experiments. Evolving analytic approaches. In H. S. Bloom (Ed.), (p. 115-172). New York: Russell Sage.

Bloom, H. S., Richburg-Hayes, L., & Black, A. R. (2007). Using covariates to improve precision for studies that randomize schools to evaluate educational interventions. *Educational Evaluation and Policy Analysis*, *29*(1), 30-59.

Bonett, D. G. (2002). Sample size requirements for estimating intraclass correlations with desired precision. *Statistics in Medicine*, *21*(9), 1331-

1335.

Boomsma, A. (2013). Reporting Monte Carlo studies in structural equation modeling. *Structural Equation Modeling*, *20*(3), 518-540.

Borm, G., van der Wilt, G., Kremer, J., & Zielhuis, G. (2007). A generalized concept of power helped to choose optimal endpoints in clinical trials. *J Clin Epidemiol*, *60*, 375-81.

Borm, G. F., Bloem, B. R., Munneke, M., & Teerenstra, S. (2010). A simple method for calculating power based on a prior trial. *Journal of Clinical Epidemiology*, *63*(9), 992-997.

Borm, G. F., Den Heijer, M., & Zielhuis, G. A. (2009). Publication bias was not a good reason to discourage trials with low power. *Journal of Clinical Epidemiology*, *62*(1), 47-53.

Borm, G. F., Hoogendoorn, E. H., Den Heijer, M., & Zielhuis, G. A. (2005). Sequential balancing: A simple method for treatment allocation in clinical trials. *Contemporary Clinical Trials*, *26*(6), 637-645.

Borm, G. F., Houben, R. M. G. J., Welsing, P. M. J., & Zielhuis, G. A. (2006). An investigation of clinical studies suggests those with multiple objectives should have at least 90% power for each endpoint. *Journal of Clinical Epidemiology*, *59*(1), 1-6.

Borm, G. F., Melis, R. J. F., Teerenstra, S., & Peer, P. G. (2005). Pseudo cluster randomisation: a treatment allocation method to minimize contamination and selection bias. *Statistics in Medicine*, *24*(23), 3535-3547.

Bosker, R., Snijders, T., & Guldemond, H. (2003). *PINT user's manual, Version 2.1.*

Brandon, P. R., Harrison, G. M., & Lawton, B. E. (2012). SAS code for calculating intraclass correlation coefficients and effect size benchmarks for site-randomized education experiments. *American Journal of Evaluation*, *34*(1), 85-90.

Broberg, P. (2013). Sample size re-assessment leading to a raised sample size does not inflate type I error rate under mild conditions. *BMC Medical Research Methodology*, *13*, 94.

Brown, B. W. (1980). The crossover experiment for clinical trials. *Biometrics*, *36*(1), 69-79.

Brown, C. A., & Lilford, R. J. (2006). The stepped wedge trial design: a systematic review. *BMC Medical Research Methodology*, *6*, 54.

Brown, H., & Prescott, R. (2015). *Applied mixed models in medicine.* Chichester: Wiley.

Browne, R. H. (1995). On the use of a pilot sample for sample size determination. *Statistics in Medicine*, *14*(17), 1933-1940.

Browne, W. J., Golalizadeh Lahi, M., & Parker, R. M. A. (2009). *A guide to sample size calculations for random effect models via simulation and the MLPowSim software package.* Bristol: School of Clinical Veterinary Sciences, University of Bristol.

Burton, A., Altman, D. G., Royston, P., & Holder, R. L. (2006). The design of simulation studies in medical statistics. *Statistics in Medicine*, *25*(24),

4279-4292.

Campbell, M. J., Donner, A., & Klar, N. (2007). Developments in cluster randomized trials and statistics in medicine. *Statistics in Medicine*, *26*(1), 2-19.

Campbell, M. J., & Walters, S. J. (2014). *How to design, analyse and report cluster randomised trials in medicine and health related research.* Chichester: Wiley.

Campbell, M. K., Elbourne, D. R., & Altman, D. G. (2004). CONSORT statement: extension to cluster randomised trials. *British Medical Journal*, *328*, 702-708.

Campbell, M. K., Fayers, P. M., & Grimshaw, J. M. (2005). Determinants of the intracluster correlation coefficient in cluster randomized trials: the case of implementation research. *Clinical Trials*, *2*(2), 99-107.

Campbell, M. K., Grimshaw, J. M., & Elbourne, D. R. (2004). Intracluster correlation coefficients in cluster randomized trials: empirical insights into how should they be reported. *BMC Medical Research Methodology*, *4*, 9.

Campbell, M. K., Mollison, J., & Grimshaw, J. M. (2001). Cluster trials in implementation research: estimation of intracluster correlation coefficients and sample size. *Statistics in Medicine*, *20*(3), 391-399.

Candel, M. J. J. M. (2009). Optimal designs for empirical Bayes estimators of individual linear and quadratic growth curves in linear mixed models. *Statistical Methods in Medical Research*, *18*(4), 397-419.

Candel, M. J. J. M., & Van Breukelen, G. J. P. (2010). Sample size adjustments for varying cluster sizes in cluster randomized trials with binary outcomes analyzed with second-order PQL mixed logistic regression. *Statistics in Medicine*, *29*(14), 1488-1501.

Carlin, J. B., & Hocking, J. (1999). Design of cross-sectional surveys using cluster sampling: an overview with Australian case studies. *Australian and New Zealand Journal of Public Health*, *23*(5), 546–551.

Carter, B. R., & Hood, K. (2008). Balance algorithm for cluster randomized trials. *BMC Medical Research Methodology*, *8*, 65.

Charles, P., Giraudeau, B., Dechartres, A., Baron, G., & Ravaud, P. (2009). Reporting of sample size calculation in randomised controlled trials: review. *British Medical Journal*, *338*, b1732.

Chow, S.-C., Shao, J., & Wang, H. (2008). *Sample size calculations in clinical research* (2nd. ed.). New York: Chapman & Hall.

Cohen, J. (1962). The statistical power of abnormal social psychological research: A review. *Journal of Abnormal and Social Psychology*, *65*(3), 145-153.

Cohen, J. (1988). *Statistical power analysis for the behavioral sciences* (2nd. ed.). New Jersey: Erlbaum.

Cools, W. (2008). *Multilevel design efficiency using simulation* (Unpublished doctoral dissertation). Catholic University Leuven.

Cornfield, J. (1978). Randomization by group: a formal analysis. *American Journal of Epidemiology, 108*(2), 100-102.

Cousineau, D. (2007). Computing the power of a t test. *Tutorials in Quantitative Methods for Psychology, 3*(2), 60-62.

Craven, R. G., Marsh, H. W., Debus, R. L., & Jayasinghe, U. (2001). Diffusion effects: control group contamination threats to the validity of teacher-administered interventions. *Journal of Educational Psychology, 93*(3), 639-645.

Crespi, C. M., Maxwell, A. E., & Wu, S. (2011). Cluster randomized trials of cancer screening interventions: Are appropriate statistical methods being used? *Contemporary Clinical Trials, 32*(4), 477-484.

Crits-Christoph, P., & Mintz, J. (1991). Implications of therapist effects for the design and analysis of comparative studies of psychotherapists. *Journal of Consulting and Clinical Psychology, 59*(1), 20-26.

Dattalo, P. (2008). *Determining sample size: Balancing power, precision, and practicality.* Oxford: Oxford University Press.

De Hoop, E., Moerbeek, M., Borsje, P., & Teerenstra, S. (2014). *Sample size estimation in cluster randomized trials with three continuous outcome measurements: the extended ANCOVA versus the stepped wedge design.* (Manuscript submitted for publication)

De Hoop, E., Teerenstra, S., Van Gaal, B. G., Moerbeek, M., & Borm, G. (2012). The "best balance" allocation led to optimal balance in cluster-controlled trials. *Journal of Clinical Epidemiology, 65*(2), 132-137.

De Jong, K., Moerbeek, M., & Van Der Leeden, R. (2010). A priori power analysis in longitudinal three-level multilevel models: an example with therapist effects. *Psychotherapy Research, 20*(3), 273-284.

Dempster, A. P., Rubin, D. B., & Tsutakawa, R. K. (1981). Estimation in covariance components models. *Journal of the American Statistical Association, 76*(374), 341-353.

Descôteaux, J. (2007). Statistical power: an historical introduction. *Tutorials in Quantitative Methods for Psychology, 3*(2), 28-34.

De Smet, A. M. G. A., Kluytmans, J. A. J. W., Cooper, B. S., Mascini, E. M., Benus, R. F. J., Van Der Werf, T. S., ... Bonten, M. J. M. (2009). Decontamination of the digestive tract and oropharynx in ICU patients. *New England Journal of Medicine, 360*(1), 20-31.

De Vaus, D. (2005). *Research design in social research.* London: Sage.

Diehr, P., Martin, D. C., Koepsell, T., & Cheadle, A. (1995). Breaking the matches in a paired t-test for community interventions when the number of pairs is small. *Statistics in Medicine, 14*, 1491-1504.

Diez Roux, A. V. (2002). A glossary for multilevel analysis. *Journal of Epidemiology and Community Health, 56*(8), 588-594.

DiSantostefano, R. L., & Muller, K. E. (1995). A comparison of power approximations for Satterthwaite's test. *Communications in Statistics, Simulation and Computation, 24*(3), 583-593.

Dolbeault, S., Cayrou, S., Bredart, A., Viala, A. L., Desclaux, B., Saltel, P.,

... Dickes, P. (2009). The effectiveness of a psycho-educational group after early-stage breast cancer treatment: results of a randomized French study. *Psycho-Oncology*, *18*(6), 647-656.

Donner, A., Birkett, N., & Buck, C. (1981). Randomization by cluster: sample size requirements and analysis. *American Journal of Epidemiology*, *114*(6), 906-914.

Donner, A., & Klar, N. (2000a). Cluster randomization trials. *Statistical Methods in Medical Research*, *9*(2), 79-80.

Donner, A., & Klar, N. (2000b). *Design and analysis of cluster randomization trials in health research.* London: Edward Arnold.

Donner, A., Klar, N., & Zou, G. (2004). Methods for the statistical analysis of binary data in split-cluster designs. *Biometrics*, *60*(4), 919-925.

Donner, A., & Koval, J. J. (1983). A note on the accuracy of Fisher's approximation to the large- sample variance of an intraclass correlation. *Communications in Statistics, Simulation and Computation*, *12*(4), 443-449.

Donner, A., & Wells, G. (1986). A comparison of confidence interval methods for the intraclass correlation coefficient. *Biometrics*, *42*(2), 401-412.

Dumville, J. C., Hahn, S., Miles, J. N. V., & Torgerson, D. J. (2006). The use of unequal randomisation ratios in clinical trials: A review. *Contemporary Clinical Trials*, *27*(1), 1-12.

Duncan, T. E., Duncan, S. C., & Strycker, L. A. (2006). *An introduction to latent variable growth curve modeling* (2nd ed.). Mahwah, NJ: Erlbaum.

Dyas, J. V., Apekey, T., Tilling, M., & Siriwardena, A. N. (2009). Strategies for improving patient recruitment to focus groups in primary care: a case study reflective paper using an analytical framework. *BMC Medical Research Methodology*, *9*, 65.

Elashoff, J. D. (2007). *nQuery Advisor Version 7.0 User's Guide.* Cork: Statistical Solutions.

Eldridge, S., Cryer, C., Feder, G., & Underwood, M. (2001). Sample size calculations for intervention trials in primary care randomizing by primary care group: an empirical illustration from one proposed intervention trial. *Statistics in Medicine*, *20*(3), 367-376.

Eldridge, S., & Kerry, S. (2012). *A practical guide to cluster randomised trials in health services research.* Chichester: Wiley.

Eldridge, S. M., Ashby, D., Feder, G. S., Rudnicka, A. R., & Ukoumunne, O. C. (2004). Lessons for cluster randomized trials in the twenty-first century: a systematic review of trials in primary care. *Clinical Trials*, *1*(1), 80-90.

Elley, C. R., Kerse, N., Chondros, P., & Robinson, E. (2005). Intraclass correlation coefficients from three cluster randomised controlled trials in primary and residential health care. *Australian and New Zealand Journal of Public Health*, *29*(5), 461–467.

Fan, X. (2003). Power of latent growth modeling for detecting group differences in linear growth trajectory parameters. *Structural Equation*

Modeling, *10*(3), 380-400.

Fan, X., & Fan, X. (2005). Power of latent growth modeling for detecting linear growth: number of measurements and comparison with other analytic approaches. *The Journal of Experimental Education*, *73*(2), 121-139.

Farrin, A., Russell, I., Torgerson, D., & Underwood, M. (2005). Differential recruitment in a cluster randomized trial in primary care: the experience of the UK Back Pain, Exercise and Active Management and Manipulation (UK BEAM) Feasibility Study. *Clinical Trials*, *2*(2), 119-124.

Faul, F., Erdfelder, E., Lang, A.-G., & Buchner, A. (2007). G*Power 3: A flexible statistical power analysis for the social, behavioral, and biomedical sciences. *Behavior Research Methods*, *39*(2), 175-191.

Feder, G., Griffiths, C., Eldridge, S., & Spence, M. (1999). Effect of postal prompts to patients and general practitioners on the quality of primary care after a coronary event (POST): randomised controlled trial. *British Medical Journal*, *318*, 1522-1526.

Feng, Z., Diehr, P., Yasui, Y., Evans, B., Beresford, S., & Koepsell, T. D. (1999). Explaining community-level variance in group randomized trials. *Statistics in Medicine*, *18*(5), 539-556.

Field, A. (2013). *Discovering statistics using IBM SPSS statistics* (4th ed.). Los Angeles: Sage.

Fisher, R. A. (1926). The arrangement of field experiments. *Journal of the Ministry of Agriculture of Great Britain*, *33*, 503-513.

Fisher, R. A. (1935). *The design of experiments.* Edinburgh: Oliver & Boyd.

Flynn, T. N., Whitley, E., & Peters, T. J. (2002). Recruitment strategies in a cluster randomized trial - cost implications. *Statistics in Medicine*, *21*(3), 397-405.

Forsetlund, L., Chalmers, I., & Bjørndal, A. (2007). When was random allocation first used to generate comparison groups in experiments to assess the effects of social interventions? *Economics of Innovation and New Technology*, *16*(5), 371-384.

Friede, T., & Kieser, M. (2006). Sample size recalculation in internal pilot study designs: a review. *Biometrical Journal*, *48*(4), 537-555.

Friedman, L. M., Furberg, C. D., & DeMets, D. L. (1999). *Fundamentals of clinical trials* (3rd ed.). New York: Springer.

Gail, M. H., Mark, S. D., Carroll, R. J., & Green, S. B. (1996). On design considerations and randomization-based inference for community intervention trials. *Statistics in Medicine*, *15*(11), 1069-1092.

Galbraith, S., & Marschner, I. C. (2002). Guidelines for the design of clinical trials with longitudinal outcomes. *Controlled Clinical Trials*, *23*(3), 257-273.

Gelman, A., & Hill, J. (2007). *Data analysis using regression and multilevel/hierarchical models.* Cambridge: Cambridge University Press.

Gilbody, S., Bower, P., Torgerson, D., & Richards, D. (2008). Cluster randomized trials produced similar results to individually randomized trials in a meta-analysis of enhanced care for depression. *Journal of Clinical*

Epidemiology, *61*(1), 160-168.

Giraudeau, B., & Ravaud, P. (2009). Preventing bias in cluster randomised trials. *PLOS Medicine*, *6*(5), 1-6.

Giraudeau, B., Ravaud, P., & Donner, A. (2008). Sample size calculation for cluster randomized cross-over trials. *Statistics in Medicine*, *27*(27), 5578-5585.

Goldstein, H. (1986). Multilevel mixed linear model analysis using iterative generalized least squares. *Biometrika*, *73*(1), 43-56.

Goldstein, H. (1989). Restricted unbiased iterative generalized least squares estimation. *Biometrika*, *76*(3), 622-623.

Goldstein, H. (1991). Nonlinear multilevel models, with an application to discrete response data. *Biometrika*, *78*(1), 45-51.

Goldstein, H. (2011). *Multilevel statistical models* (4th ed.). Chichester: Wiley.

Goldstein, H., & Rasbash, J. (1996). Improved approximation for multilevel models with binary responses. *Journal of the Royal Statistical Society Series A*, *159*(3), 505-513.

Goos, P., & Jones, B. (2011). *Optimal design of experiments. a case study approach.* Chichester: Wiley.

Gould, A. L. (1998). Multi-centre trial analysis revisited. *Statistics in Medicine*, *17*(15-16), 1779-1797.

Grissom, R. J., & Kim, J. J. (2005). *Effect sizes for research.* Mahwah, NJ: Erlbaum.

Guittet, L., Giraudeau, B., & Ravaud, P. (2005). A priori postulated and real power in cluster randomized trials: mind the gap. *BMC Medical Research Methodology*, *5*, 25.

Guittet, L., Ravaud, P., & Giraudeau, B. (2006). Planning a cluster randomized trial with unequal cluster sizes: practical issues involving continuous outcomes. *BMC Medical Research Methodology*, *6*, 17.

Gulliford, M. C., Adams, G., Ukoumunne, O. C., Latinovic, R., Chinn, S., & Campbell, M. J. (2005). Intraclass correlation coefficient and outcome prevalence are associated in clustered binary data. *Journal of Clinical Epidemiology*, *58*(3), 246-251.

Gulliford, M. C., Ukoumunne, O. C., & Chinn, S. (1999). Components of variance and intraclass correlations for the design of community-based surveys and intervention studies: data from the health survey for England 1994. *American Journal of Epidemiology*, *149*(9), 876-883.

Haddad, S. M., Sousa, M. H., Cecatti, J. G., Parpinelli, M. A., Costa, M. L., & Souza, J. (2012). Intraclass correlation coefficients in the Brazilian network for surveillance of severe maternal morbidity study. *BMC Pregnancy and Childbirth*, *12*, 101.

Hahn, S., Puffer, S., Torgerson, D. J., & Watson, J. (2005). Methodological bias in cluster randomised trials. *BMC Medical Research Methodology*, *5*, 10.

Hannan, P., Murray, D., Jacobs, D., & McGovern Jr., P. (1994). Parameters to aid in the design and analysis of community trials: intraclass correlations from the Minnesota Heart Health Program. *Epidemiology*, *5*(1), 88-95.

Harrison, D. A., & Brady, A. R. (2004). Sample size and power calculations using the noncentral *t*-distribution. *The Stata Journal*, *4*(2), 142-153.

Hartley, H. O., & Rao, J. N. K. (1967). Maximum-likelihood estimation for the mixed analysis of variance model. *Biometrika*, *54*(1), 93-108.

Hawkins, J. D., Van Horn, M. L., & Arthur, M. W. (2004). Community variation in risk and protective factors and substance use outcomes. *Prevention Science*, *5*(4), 213–220.

Hayes, R. J., & Bennett, S. (1999). Simple sample size calculation for cluster-randomized trials. *International Journal of Epidemiology*, *28*(2), 319-326.

Hayes, R. J., & Moulton, L. H. (2009). *Cluster randomised trials.* Boca Raton: CRC Press.

Headrick, T. C., & Zumbo, B. D. (2005). On optimizing multi-level designs: power under budget constraints. *Australian and New Zealand Journal of Statistics*, *47*(2), 219-229.

Heck, R. H., & Thomas, S. L. (2015). *An introduction to multilevel modeling techniques: Mlm and sem approaches using mplus* (3rd ed.). New York: Routledge.

Hedberg, E., & Hedges, L. V. (2014). Reference values of within-district intraclass correlations of academic achievement by district characteristics: results from a meta-analysis of district-specific values. *Evaluation Review*, *38*(6), 546-582.

Hedeker, D., Gibbons, R., du Toit, M., & Chen, Y. (2008). *Supermix. Mixed effects models.* Lincolnwood, IL: Scientific Software International.

Hedeker, D., & Gibbons, R. D. (1994). A random-effects ordinal regression model for multilevel analysis. *Biometrics*, *50*(4), 933-944.

Hedeker, D., & Gibbons, R. D. (2006). *Longitudinal data analysis.* Hoboken: Wiley.

Hedges, L. V., & Hedberg, E. C. (2007). Intraclass correlation values for planning group-randomized trials in education. *Educational Evaluation and Policy Analysis*, *29*(1), 60–87.

Hedges, L. V., & Hedberg, E. C. (2014). Intraclass correlations and covariate outcome correlations for planning two- and three-level cluster-randomized experiments in education. *Evaluation Review*, *37*(6), 445-489.

Hemming, K., Girling, A. J., Sitch, A. J., Marsh, J., & Lilford, R. J. (2011). Sample size calculations for cluster randomisation controlled trials with a fixed number of clusters. *BMC Medical Research Methodology*, *11*, 102.

Hemming, K., Lilford, R., & Girling, A. J. (2015). Stepped-wedge cluster randomised controlled trials: a generic framework including parallel and multiplelevel designs. *Statistics in Medicine*, *34*(2), 181-196.

Hendriks, J. C. M., Teerenstra, S., Punt-Van Der Zalm, J. P. E., Wetzels,

A. M. M., Westphal, J. R., & Borm, G. F. (2005). Sample size calculations for a split-cluster, beta-binomial design in the assessment of toxicity. *Statistics in Medicine*, *24*, 3757-3772.

Heo, M., & Leon, A. C. (2008). Statistical power and sample size requirements for three level hierarchical cluster randomized trials. *Biometrics*, *64*(4), 1256-1262.

Heo, M., & Leon, A. C. (2009). Sample size requirements to detect an intervention by time interaction in longitudinal cluster randomized clinical trials. *Statistics in Medicine*, *28*(6), 1017 - 1027.

Heo, M., Litwin, A., Blackstock, O., Kim, N., & Arnsten, J. (in press). Sample size determinations for group-based randomized clinical trials with different levels of data hierarchy between experimental and control arms. *Statistical Methods in Medical Research.*

Hewitt, C. E., Torgerson, D. J., & Miles, J. N. V. (2008). Individual allocation had an advantage over cluster randomization in statistical efficiency in some circumstances. *Journal of Clinical Epidemiology*, *61*(10), 1004-1008.

Hintze, J. L. (2008). *PASS. User's Guide.* Kaysville: NCSS.

Hoddinott, P., Britten, J., Harrild, K., & Godden, D. J. (2007). Recruitment issues when primary care population clusters are used in randomised controlled clinical trials: climbing mountains or pushing boulders uphill? *Contemporary Clinical Trials*, *28*(3), 232-241.

Hoenig, J., & Heisey, D. (2001). The abuse of power: the pervasive fallacy of power calculations for data analysis. *The American Statistician*, *55*(1), 19-24.

Hoover, D. R. (2002). Clinical trials of behavioural interventions with heterogeneous teaching subgroup effects. *Statistics in Medicine*, *21*(10), 1351-1364.

Hox, J., Van De Schoot, R., & Matthijsse, S. (2012). How few countries will do? Comparative survey analysis from a Bayesian perspective. *Survey Research Methods*, *6*(2), 87-83.

Hox, J. J. (2010). *Multilevel analysis. techniques and applications* (2nd ed.). New York: Routledge.

Hox, J. J., Moerbeek, M., Kluytmans, A., & Van De Schoot, R. (2014). Analyzing indirect effects in cluster randomized trials: the effect of estimation method, number of groups and group sizes on accuracy and power. *Frontiers in Psychology*, *5*, 78.

Hsu, L. M. (1994). Unbalanced designs to maximize statistical power in psychotherapy efficacy studies. *Psychotherapy Research*, *4*(2), 95-106.

Hughes, J. P. (2005). Using baseline data to design a group randomized trial. *Statistics in Medicine*, *24*(13), 1983-1994.

Hussey, M. A., & Hughes, J. P. (2007). Design and analysis of stepped wedge cluster randomized trials. *Contemporary Clinical Trials*, *28*(2), 182-191.

Hutchison, D. (2009). Designing your sample efficiently: clustering effects in education surveys. *Educational Research*, *51*(1), 119–126.

Ip, E. H., Wasserman, R., & Barkin, S. (2011). Comparison of intraclass correlation coefficient estimates and standard errors between using cross-sectional and repeated measurement data: the safety check cluster randomized trial. *Contemporary Clinical Trials*, *32*, 225-232.

Ivers, N. M., Taljaard, M., Dixon, S., Bennett, C., McRae, A., Taleban, J., ... Donner, A. (2011). Impact of CONSORT extension for cluster randomised trials on quality of reporting and study methodology: review of random sample of 300 trials, 2000-8. *British Medical Journal*, *343*.

Jacob, R., Zhu, P., & Bloom, H. S. (2010). New empirical evidence for the design of group randomized trials in education. *Journal of Research on Educational Effectiveness*, *3*, 157-198.

Jadad, A. R., & Enkin, M. (2007). *Randomized controlled trials: Questions, answers and musings.* Malden, MA: Blackwell.

Janega, J. B., Murray, D. M., Varnell, S. P., Blitstein, J. L., Birnbaum, A. S., & Lytle, L. A. (2004a). Assessing intervention effects in a school-based nutrition intervention trial: which analytic model is most powerful? *Health Education & Behavior*, *31*(6), 756-774.

Janega, J. B., Murray, D. M., Varnell, S. P., Blitstein, J. L., Birnbaum, A. S., & Lytle, L. A. (2004b). Assessing the most powerful analysis method for school-based intervention studies with alcohol, tobacco, and other drug outcomes. *Addictive Behaviors*, *29*(3), 595-606.

Janjua, N. Z., Khan, M. I., & Clemens, J. D. (2006). Estimates of intraclass correlation coefficient and design effect for surveys and cluster randomized trials on injection use in Pakistan and developing countries. *Tropical Medicine & International Health*, *11*(12), 1832-1840.

Jensen, K., & Kieser, M. (2010). Blinded sample size recalculation in multicentre trials with normally distributed outcome. *Biometrical Journal*, *52*(3), 377-399.

Jones, B. (2005). The design and analysis of multicentre clinical trials. *Statistical Methods in Medical Research*, *14*(2), 203-204.

Jones, B., & Kenward, M. G. (2002). *Design and analysis of cross-over trials.* Boca Raton: Chapman & Hall/CRC.

Jones, B., Teather, D., Wang, J., & Lewis, J. A. (1998). A comparison of various estimators of a treatment difference for a multi-centre clinical trial. *Statistics in Medicine*, *17*(15-16), 1767-1777.

Jordhoy, M. S., Fayers, P. M., Ahlner-Elmqvist, M., & Kaasa, S. (2002). Lack of concealment may lead to selection bias in cluster randomized trials of palliative care. *Palliative Medicine*, *16*(1), 43-49.

Julious, S. A. (2010). *Sample sizes for clinical trials.* Boca Raton: Chapman & Hall/CRC Press.

Kelcey, B., & Phelps, G. (2013). Considerations for designing group randomized trials of professional development with teacher knowledge outcomes. *Educational Evaluation and Policy Analysis*, *35*(3), 370-390.

Kelcey, B., & Phelps, G. (2014). Strategies for improving power in school-randomized studies of professional development. *Evaluation Review*,

37(6), 520-554.

Keogh-Brown, M., Bachmann, M., Shepstone, L., Hewitt, C., Howe, A., Ramsay, C., ... Campbell, M. (2007). Contamination in trials of educational interventions. *Health Technology Assessment, 11*(43).

Kerry, S. M., Cappuccio, F. P., Emmett, L., Plange-Rhule, J., & Eastwood, J. B. (2005). Reducing selection bias in a cluster randomized trial in West African villages. *Clinical Trials, 2*(2), 125-129.

Kim, Y., Choi, Y.-K., & Emery, S. (2013). Logistic regression with multiple random effects: a simulation study of estimation methods and statistical packages. *The American Statistician, 67*(3), 171-182.

Kish, L. (1965). *Survey sampling.* New York: Wiley.

Klar, N., & Donner, A. (1997). The merits of matching in community intervention trials: a cautionary tale. *Statistics in Medicine, 16*, 1753-1765.

Klar, N., & Donner, A. (2001). Current and future challenges in the design and analysis of cluster randomization trials. *Statistics in Medicine, 20*(24), 3729-3740.

Knox, S. A., & Chondros, P. (2004). Observed intra-cluster correlation coefficients in a cluster survey sample of patient encounters in general practice in Australia. *BMC Medical Research Methodology, 4*, 30.

Konstantopoulos, S. (2009). Incorporating costs in power analysis for three-level cluster randomized designs. *Evaluation Review, 33*(4), 335-357.

Konstantopoulos, S. (2011). Optimal sampling of units in three-level cluster randomized designs: An ANCOVA framework. *Educational and Psychological Measurement, 71*(5), 898-813.

Konstantopoulos, S. (2012). The impact of covariates on statistical power in cluster randomized designs: Which level matters more? *Multivariate Behavioral Research, 47*(3), 392-420.

Korendijk, E. J. H. (2012). *Robustness and optimal design issues for cluster randomized trials* (Unpublished doctoral dissertation). Utrecht University.

Korendijk, E. J. H., Moerbeek, M., & Maas, C. J. M. (2010). The robustness of designs for trials with nested data against incorrect initial intracluster correlation coefficient estimates. *Journal of Educational and Behavioral Statistics, 35*(5), 566-585.

Lajos, G. J., Haddad, S. M., Tedesco, R. P., Passini Jr, R., Dias, T. Z., Nomura, M. L., ... Sousa, J. G. f., M. F. Cecatti (2014). Intracluster correlation coefficients for the Brazilian multicenter study on preterm birth (EMIP): methodological and practical implications. *BMC Medical Research Methodology, 14*, 54.

Lake, S., Kammann, E., Klar, N., & Betensky, R. A. (2002). Sample size re-estimation in cluster randomization trials. *Statistics in Medicine, 21*(10), 1337-1350.

Landau, S., & Everitt, B. S. (2004). *A handbook of statistical analyses using SPSS.* Boca Raton: Chapman & Hall/CRC.

Landau, S., & Stahl, D. (2013). Sample size and power calculations for medical studies by simulation when closed form expressions are not available. *Statistical Methods in Medical Research*, *22*(3), 324-345.

Lee, K. J., & Thompson, S. G. (2005). The use of random effects models to allow for clustering in individually randomized trials. *Clinical Trials*, *2*(2), 163-173.

Lemme, F., Van Breukelen, G. J. P., & Berger, M. P. F. (in press). Efficient treatment allocation in two-way nested designs. *Statistical Methods in Medical Research.*

Leontjevas, R., Gerritsen, D. L., Smalbrugge, M., Teerenstra, S., Vernooij-Dassen, M. J., & Koopmans, R. T. (2013). A structural multidisciplinary approach to depression management in nursing-home residents: a multicentre, stepped-wedge cluster-randomised trial. *Lancet*, *381*, 2255-64.

Lewsey, J. D. (2004). Comparing completely and stratified randomized designs in cluster randomized trials when the stratifying factor is cluster size: a simulation study. *Statistics in Medicine*, *23*(6), 897-905.

Li, B., Lingsma, H. F., Steyerberg, E. W., & Lesaffre, E. (2011). Logistic random effects regression models: a comparison of statistical packages for binary and ordinal outcomes. *BMC Medical Research Methodology*, *11*, 77.

Lin, S.-J., Tu, Y.-L., Tsai, S. C., Lai, S.-M., & Lu, H. K. (2012). Non-surgical periodontal therapy with and without subgingival minocycline administration in patients with poorly controlled type II diabetes: a randomized controlled clinical trial. *Clinical Oral Investigations*, *16*(2), 599-609.

Lindström, D., Sundberg-Petersson, I., Adami, J., & Tönnesen, H. (2010). Disappointment and drop-out rate after being allocated to control group in a smoking cessation trial. *Contemporary Clinical Trials*, *31*(1), 22-26.

Lipsey, M. W., & Wilson, D. B. (1993). The efficacy of psychological, educational, and behavioral treatment: confirmation from meta-analysis. *American Psychologist*, *48*(12), 1181-1209.

Littell, R. C., Milliken, G. A., Stroup, W. W., Wolfinger, R. D., & Schabenberger, O. (2006). *SAS for mixed models.* Cary: SAS Institute.

Littenberg, B., & MacLean, C. D. (2006). Intra-cluster correlation coefficients in adults with diabetes in primary care practices: the Vermont Diabetes Information System field survey. *BMC Medical Research Methodology*, *6*, 20.

Liu, X. F. (2003). Statistical power and optimum sample allocation ratio for treatment and control having unequal costs per unit of randomization. *Journal of Educational and Behavioral Statistics*, *28*(3), 231-248.

Liu, X. F. (2014). *Statistical power analysis for the social and behavioral sciences.* New York: Routledge.

Longford, N. T. (1987). A fast scoring algorithm for maximum likelihood estimation in unbalanced mixed models with nested random effects. *Biometrika*, *74*(4), 817-827.

Longford, N. T. (1993). *Random coefficient models.* New York: Oxford University Press.

Lyon, G. J., Samar, S. M., Conelea, C., Trujillo, M. R., Lipinski, C. M., Bauer, C. C., ... Coffey, B. J. (2010). Testing tic suppression: comparing the effects of dexmethylphenidate to no medication in children and adolescents with attention-deficit/hyperactivity disorder and Tourette's disorder. *Journal of Child and Adolescent Psychopharmacology*, *20*(4), 283-289.

Maas, C. J. M., & Hox, J. J. (2005). Sufficient sample sizes for multilevel modeling. *Methodology*, *1*(3), 86-92.

Machin, D., Campbell, M. J., Tan, S.-B., & Tan, S.-H. (2009). *Sample size tables for clinical studies* (3rd ed.). Chichester: Blackwell.

Manatunga, A. K., Hudges, M. G., & Chen, S. (2001). Sample size estimation in cluster randomized studies with varying cluster size. *Biometrical Journal*, *43*(1), 75-86.

Martin, D. C., Diehr, P., Perrin, E. B., & Koepsell, T. D. (1993). The effect of matching on the power of randomized community intervention studies. *Statistics in Medicine*, *12*(3), 329-338.

Martindale, C. (1978). The therapist-as-fixed-effect fallacy in psychotherapy research. *Journal of Consulting and Clinical Psychology*, *46*(6), 1526-1530.

Martinson, B. C., Murray, D. M., Jeffery, R. W., & Hennrikus, D. J. (1999). Intraclass correlation for measures from a worksite health promotion study: Estimates, correlates, and applications. *American Journal of Health Promotion*, *13*(3), 347-357.

Mason, W. M., Wong, G. Y., & Entwisle, B. (1983). Contextual analysis through the multilevel linear model. In S. Leinhardt (Ed.), *Sociological methodology: 1983-1984* (p. 72-103). San Francisco: Jossey-Bass.

Mayr, S., Erdfelder, E., Buchner, A., & Faul, F. (2007). A short tutorial of G*Power. *Tutorials in Quantitative Methods for Psychology*, *3*(2), 51-59.

McCullagh, P., & Nelder, J. A. (1989). *Generalized linear models.* London: Chapman & Hall.

McDonald, R., Jouriles, E. N., & Skopp, N. A. (2006). Reducing conduct problems among children brought to women's shelters: intervention effects 24 months following termination of services. *Journal of Family Psychology*, *20*(1), 127-136.

McNair, D. M., Lorr, M., & Droppleman, L. P. (1992). *EdITS Manual for the profile of mood states.* San Diego, CA: Educational and Industrial Testing Service.

Mdege, N. D., Man, M. S., Taylor nee Brown, C. A., & Torgerson, D. J. (2011). Systematic review of stepped wedge cluster randomized trials shows that design is particularly used to evaluate interventions during routine implementation. *Journal of Clinical Epidemiology*, *64*(9), 936-48.

Medical Research Council. (1948). Streptomycin treatment of pulmonary

tuberculosis. *British Medical Journal*, *2*, 782.
Melis, R. J. F., Teerenstra, S., Olde Rikkert, M. G. M., & Borm, G. (2011). Pseudo cluster randomization: balancing the disadvantages of cluster and individual randomization. *Evaluation and the Health Professions*, *34*(2), 151-163.
Melis, R. J. F., Teerenstra, S., Olde Rikkert, M. G. M., & Borm, G. F. (2008). Pseudo cluster randomization performed well when used in practice. *Journal of Clinical Epidemiology*, *61*(11), 1169-1175.
Melis, R. J. F., Van Eijken, M. I. J., & Borm, G. F. (2005). The design of the Dutch EASYcare study: a randomised controlled trial on the effectiveness of a problem-based community intervention model for frail elderly people. *BMC Health Services Research*, *5*(1), 65.
Metcalf, P. A., Scragg, R. K. R., Stewart, A. W., & Scott, A. J. (2007). Design effects associated with dietary nutrient intakes from a clustered design of 1- to 14-year-old children. *European Journal of Clinical Nutrition*, *61*(1), 1064-1071.
Moerbeek, M. (2004). The consequence of ignoring a level of nesting in multilevel analysis. *Multivariate Behavioral Research*, *39*(1), 129-149.
Moerbeek, M. (2005). Randomization of clusters versus randomization of persons within clusters: which is preferable? *The American Statistician*, *59*(1), 72-78.
Moerbeek, M. (2006a). Cluster randomized trials: design and analysis. In H. Pham (Ed.), *Handbook of engineering statistics* (p. 705-718). London: Springer.
Moerbeek, M. (2006b). Power and money in cluster randomized trials: when is it worth measuring a covariate? *Statistics in Medicine*, *25*(15), 2607-2617.
Moerbeek, M. (2008). Powerful and cost-efficient designs for longitudinal intervention studies with two treatment groups. *Journal of Educational and Behavioral Statistics*, *33*(1), 41-61.
Moerbeek, M., & Maas, C. J. M. (2005). Optimal experimental designs for multilevel logistic models with two binary predictors. *Communications in Statistics, Theory and Methods*, *34*(5), 1151-1167.
Moerbeek, M., & Teerenstra, S. (2010). Optimal design in multilevel experiments. In J. Hox & J. Roberts (Eds.), *Handbook of advanced multilevel analysis* (p. 257-281). New York: Routledge.
Moerbeek, M., Van Breukelen, G. J. P., & Berger, M. P. F. (2000). Design issues for experiments in multilevel populations. *Journal of Educational and Behavioral Statistics*, *25*(3), 271-284.
Moerbeek, M., Van Breukelen, G. J. P., & Berger, M. P. F. (2001a). Optimal experimental designs for multilevel logistic models. *The Statistician*, *50*(1), 1-14.
Moerbeek, M., Van Breukelen, G. J. P., & Berger, M. P. F. (2001b). Optimal experimental designs for multilevel models with covariates. *Communications in Statistics, Theory and Methods*, *30*(12), 2683-2697.

Moerbeek, M., Van Breukelen, G. J. P., & Berger, M. P. F. (2003a). A comparison between traditional methods and multilevel regression for the analysis of multi-center intervention studies. *Journal of Clinical Epidemiology*, *56*(4), 341-350.

Moerbeek, M., Van Breukelen, G. J. P., & Berger, M. P. F. (2003b). A comparison of estimation methods for multilevel logistic models. *Computational Statistics*, *18*(1), 19-37.

Moerbeek, M., Van Breukelen, G. J. P., & Berger, M. P. F. (2003c). Optimal sample sizes in experimental designs with individuals nested within clusters. *Understanding Statistics*, *2*(3), 151-175.

Moerbeek, M., Van Breukelen, G. J. P., & Berger, M. P. F. (2008). Optimal designs for multilevel studies. In J. De Leeuw & E. Meijer (Eds.), *Handbook of multilevel analysis* (p. 177-205). New York: Springer.

Moerbeek, M., & Wong, W. K. (2008). Sample size formulae for trials comparing group and individual treatments in a multilevel model. *Statistics in Medicine*, *27*(15), 2850-2864.

Moher, D., Hopewell, S., Schulz, K., Montori, V., Gøtzsche, P., Devereaux, P., ... Altman, D. (2010). Consort 2010 explanation and elaboration: updated guidelines for reporting parallel group randomised trials. *Journal of Clinical Epidemiology*, *63*, e1-e37.

Moher, D., Schulz, K. F., & Altman, D. G. (2001). The CONSORT statement: revised recommendations for improving the quality or reports of parallel-group randomised trials. *Lancet*, *134*, 663-94.

Moineddin, R., Matheson, F. I., & Glazieer, R. H. (2007). A simulation study of sample size for multilevel logistic regression models. *BMC Medical Research Methodology*, *7*, 34.

Mok, M. (1996). Sample size requirements for 2-level designs in educational research. *Multilevel Modelling Newsletter*, *7*, 11-16.

Moore, H., Summerbell, C. D., Vail, A., Greenwood, D. C., & Adamson, A. J. (2001). The design features and practicabilities of conducting a pragmatic cluster randomized trial of obesity management in primary care. *Statistics in Medicine*, *20*(3), 331-340.

Moser, B. K., Steves, G. R., & Watts, C. L. (1989). The two-sample t test versus Satterthwaite's approximate F test. *Communications in Statistics, Theory and Methods*, *18*(11), 3963-3975.

Moulton, L. H. (2004). Covariate-based constrained randomization of group-randomized trials. *Clinical Trials*, *1*(3), 297-305.

Moulton, L. H. (2005). A practical look at cluster-randomized trials. *Clinical Trials*, *2*(2), 89-90.

Murphy, K. R., Myors, B., & Wolach, A. (2008). *Statistical power analysis* (3rd. ed.). New York: Routledge.

Murray, D. M. (1998). *Design and analysis of group-randomized trials.* New York: Oxford University Press.

Murray, D. M. (2001). Statistical models appropriate for designs often used in group-randomized trials. *Statistics in Medicine*, *20*(9-10), 1373-1385.

Murray, D. M., Alfano, C. M., Zbikowski, S. M., Padgett, L. S., Robinson, L. A., & Klesges, R. (2002). Intraclass correlation among measures related to cigarette use by adolescents. estimates from an urban and largely African American cohort. *Addictive Behaviors*, *27*(4), 509-527.

Murray, D. M., & Blitstein, J. L. (2003). Methods to reduce the impact of intraclass correlation in group-randomized trials. *Evaluation Review*, *27*(1), 79-103.

Murray, D. M., Blitstein, J. L., Hannan, P. J., Baker, W. L., & Lytle, L. A. (2007). Sizing a trial to alter the trajectory of health behaviours: methods, parameter estimates, and their application. *Statistics in Medicine*, *26*(11), 2297-2316.

Murray, D. M., & Hannan, P. J. (1990). Planning for the appropriate analysis in school-based drug-use prevention studies. *Journal of Consulting and Clinical Psychology*, *58*(4), 458-468.

Murray, D. M., Hannan, P. J., Wolfinger, R. D., Baker, W., & Dwyer, J. H. (1998). Analysis of data from group-randomized trials with repeat observations on the same groups. *Statistics in Medicine*, *17*(14), 1581-1600.

Murray, D. M., Pals, S. L., Blitstein, J. L., Alfano, C. M., & Lehman, J. (2008). Design and analysis of group-randomized trials in cancer: A review of current practices. *Journal of the National Cancer Institute*, *100*(7), 483-491.

Murray, D. M., Phillips, G. A., Birnbaum, A. S., & Lytle, L. A. (2001). Intraclass correlation for measures from a middle school nutrition intervention study: Estimates, correlates, and applications. *Health Education & Behavior*, *28*(6), 666-679.

Murray, D. M., Rooney, B. L., Hannan, P. J., Peterson, A. V., Ary, D. V., Biglan, A., ... Schinke, S. P. (1994). Intraclass correlation among common measures of adolescent smoking: estimates, correlates, and applications in smoking prevention studies. *American Journal of Epidemiology*, *140*(11), 1038-1050.

Murray, D. M., & Short, B. (1995). Intraclass correlation among measures related to alcohol use by young adults: estimates, correlates and applications in intervention studies. *Journal of Studies On Alcohol*, *56*(6), 681–694.

Murray, D. M., & Short, B. (1996). Intraclass correlation among measures related to alcohol use by school aged adolescents: estimates, correlates and applications in intervention studies. *Journal of Drug Education*, *26*(3), 207–230.

Murray, D. M., & Short, B. J. (1997). Intraclass correlation among measures related to tobacco use by adolescents: estimates, correlates, and applications in intervention studies. *Addictive Behaviors*, *22*(1), 1-12.

Murray, D. M., Stevens, J., Hannan, P. J., Catellier, D. J., Schmitz, K. H., Dowda, M., ... Yang, S. (2006). School-level intraclass correlation for physical activity in sixth grade girls. *Medicine and Science in Sports and Exercise*, *38*(5), 926–936.

Murray, D. M., Van Horn, M., Hawkins, J. D., & Arthur, M. W. (2006). Analysis strategies for a community trial to reduce adolescent ATOD use: A comparison of random coefficient and ANOVA/ANCOVA models. *Contemporary Clinical Trials*, *27*(2), 188-206.

Murray, D. M., Varnell, S. P., & Blitstein, J. L. (2004). Design and analysis of group-randomized trials: A review of recent methodological developments. *American Journal of Public Health*, *94*(3), 423-432.

Muthén, B. O., & Muthén, L. K. (2012). *Mplus user's guide.* Los Angeles, CA: Muthén and Muthén.

Muthén, L. K., & Muthén, B. O. (2002). How to use a Monte Carlo study to decide on sample size and determine power. *Structural Equation Modeling*, *9*(2), 599-620.

Nietert, P. J., Jenkins, R. G., Nemeth, L. S., & Ornstein, S. M. (2009). An application of a modified constrained randomization process to a practice-based cluster randomized trial to improve colorectal cancer screening. *Contemporary Clinical Trials*, *30*(2), 129-132.

Onwuegbuzie, A. J., & Leech, N. L. (2004). Post hoc power: a concept whose time has come. *Understanding Statistics*, *3*(4), 201-230.

Opdenakker, M.-C., & Van Damme, J. (2000). The importance of identifying levels in multilevel analysis: an illustration of the effects of ignoring the top or intermediate levels in school effectiveness research. *School Effectiveness and School Improvement*, *11*(1), 103-130.

Otte, M. W., Pollack, M. H., Gould, R. A., Wothington, J. J., McArdle, E. T., & Rosenbaum, J. F. (2000). A comparison of the efficacy of clonazepam and cognitive-behavioral group therapy for the treatment of social phobia. *Journal of Anxiety Disorders*, *14*(4), 345-358.

Pagel, C., Prost, A., Lewycka, S., Das, S., Colbourn, T., Mahapatra, R., ... Osrin, D. (2011). Intracluster correlation coefficients and coefficients of variation for perinatal outcomes from five cluster-randomised controlled trials in low and middle-income countries: results and methodological implications. *Trials*, *12*, 151.

Pals, S. L., Beaty, B. L., Posner, S. F., & Bull, S. S. (2009). Estimates of intraclass correlation for variables related to behavioral HIV/STD prevention in a predominantly African American and Hispanic sample of young women. *Health Education & Behavior*, *36*(1), 182-194.

Pals, S. P., Murray, D. M., Alfano, C. M., Shadish, W. R., Hannan, P. J., & Baker, W. L. (2008). Individually randomized group treatment trials: A critical appraisal of frequently used design and analytic approaches. *American Journal of Public Health*, *98*(8), 1418-1424.

Parker, D. A., Evangelou, E., & Eaton, C. B. (2005). Intraclass correlation coefficients for cluster randomized trials in primary care: The cholesterol education and research trial (CEART). *Contemporary Clinical Trials*, *26*(2), 260-267.

Parzen, M., Lipsitz, S. R., & Dear, K. B. G. (1998). Does clustering affect the usual test statistics of no treatment effect in a randomized clinical trial? *Biometrical Journal*, *40*(4), 385-402.

Paxton, P., Curran, P. J., Bollen, K. A., Kirby, J., & Chen, F. (2001). Monte Carlo experiments: design and implementation. *Structural Equation Modeling*, *8*(2), 287-312.

Piaggio, G., Carroli, G., Villar, J., Pinol, A., Bakketeig, L., Lumbiganon, P., ... Berendes, H. (2001). Methodological considerations on the design and analysis of an equivalence stratified cluster randomization trial. *Statistics in Medicine*, *20*(3), 401-416.

Pinheiro, J. C., & Bates, D. M. (2004). *Mixed-effects models in S and S-PLUS.* New York: Springer.

Plewis, I., & Hurry, J. (1998). A multilevel perspective on the design and analysis of intervention studies. *Educational Research and Evaluation*, *4*(1), 13-26.

Pocock, S. J., & Simon, R. (1975). Sequential treatment assignment with balancing for prognostic factors in the controlled clinical trial. *Biometrics*, *31*(1), 103-115.

Preisser, J. S., Reboussin, B. A., Song, E.-Y., & Wolfson, M. (2007). The importance and role of intracluster correlations in planning cluster trials. *Epidemiology*, *18*(5), 552–560.

Proschan, M. (2009). Sample size re-estimation in clinical trials. *Biometrical Journal*, *51*(2), 348-357.

Puffer, S., Torgerson, D., & Watson, J. (2003). Evidence for risk of bias in cluster randomised trials: review of recent trials published in three general medical journals. *British Medical Journal*, *327*(7418), 785-789.

Rabe-Hesketh, S., & Skrondal, A. (2012a). *Multilevel and longitudinal modeling using stata.* (3rd ed., Vol. I: Continuous Responses). College Station, TX: Stata Press.

Rabe-Hesketh, S., & Skrondal, A. (2012b). *Multilevel and longitudinal modeling using stata* (3rd ed., Vols. II: Categorical Responses, Counts, and Survival). College Station, TX: Stata Press.

Rasbash, J., Steele, F., Browne, W. J., & Goldstein, H. (2012). *A user's guide to MLwiN.* Bristol: Centre for multilevel modelling.

Raudenbush, S. W. (1993). Hierarchical linear models and experimental design. In L. Edwards (Ed.), *Applied analysis of variance in behavioral science* (p. 459-496). New York: Wiley.

Raudenbush, S. W. (1997). Statistical analysis and optimal design for cluster randomized studies. *Psychological Methods*, *2*(2), 173-185.

Raudenbush, S. W., & Bryk, A. (2002). *Hierarchical linear models. Applications and data analysis methods.* Thousand Oaks: Sage Publications.

Raudenbush, S. W., Bryk, A. S., Cheong, Y. F., Congdon, R. T., & du Toit, M. (2011). *HLM 7. Hierarchical linear and nonlinear modeling.* Lincolnwood: Scientific Software International.

Raudenbush, S. W., & Liu, X. (2000). Statistical power and optimal design for multisite randomized trials. *Psychological Methods*, *5*(2), 199-213.

Raudenbush, S. W., & Liu, X. (2001). Effects of study duration, frequency of observation, and sample size on power in studies of group differences in polynomial change. *Psychological Methods*, *6*(4), 387-401.

Raudenbush, S. W., Martinez, A., & Spybrook, J. (2007). Strategies for improving precision in group-randomized experiments. *Educational Evaluation and Policy Analysis*, *29*(1), 5-29.

Raudenbush, S. W., Yang, M.-L., & Yosef, M. (2000). Maximum likelihood for generalized linear models with nested random effects via higher-order, multivariate Laplace approximation. *Journal of Computational and Graphical Statistics*, *9*(1), 141-157.

Reading, R., Harvey, I., & McLean, M. (2000). Cluster randomised trials in maternal and child health: implications for power and sample size. *Archives of Disease In Childhood*, *82*(1), 79–83.

Resnocow, K., Zhang, N., Vaughan, R. D., Reddy, S. P., James, S., & Murray, D. M. (2010). When intraclass correlation coefficients go awry: a case study from a school-based smoking prevention study in South Africa. *American Journal of Public Health*, *100*(9), 1714-1718.

Rhoads, C. H. (2011). The implications of "contamination" for experimental design in education. *Journal of Educational and Behavioral Statistics*, *36*(1), 76-104.

Rietbergen, C., & Moerbeek, M. (2011). The design of cluster randomized crossover trials. *Journal of Educational and Behavioral Statistics*, *36*(4), 472-490.

Roberts, C. (1999). The implications of variation in outcome between health professionals for the design and analysis of randomized controlled trials. *Statistics in Medicine*, *18*(19), 2605 - 2615.

Roberts, C., & Roberts, S. A. (2005). The design and analysis of clinical trials with clustering effects due to treatment. *Clinical Trials*, *2*(2), 152-162.

Rodríguez, G., & Goldman, N. (1995). An assessment of estimation procedures for multilevel models with binary data. *Journal of the Royal Statistical Society Series A*, *158*(1), 73-89.

Rodríguez, G., & Goldman, N. (2001). Improved estimation procedures for multilevel models with binary response: A case-study. *Journal of the Royal Statistical Society Series A*, *164*(2), 339-355.

Röhmel, J. (2006). Editorial: Adaptive designs: expectations are high. *Biometrical Journal*, *48*(4), 491-492.

Ronan, G., Gerhart, J. I., Dollard, K., & Maurelli, K. A. (2010). An analysis of survival time to re-arrest in treated and non-treated jailers. *The Journal of Forensic Psychiatry & Psychology*, *21*(1), 102-112.

Rotondi, M. A., & Donner, A. (2009). Sample size estimation in cluster randomized educational trials: an empirical Bayes approach. *Journal of Educational and Behavioral Statistics*, *34*(2), 229-237.

Roudsari, B., Fowler, R., & Nathens, A. (2007). Intracluster correlation coefficient in multicenter childhood trauma studies. *Injury Prevention*, *13*(5), 344–347.

Roudsari, B., Nathens, A., Koepsell, T., Mock, C., & Rivara, F. (2006). Analysis of clustered data in multicentre trauma studies. *Injury: International Journal of the Care of the Injured*, *37*(7), 614–621.

Rowe, A. K., Lama, M., Onikpo, F., & Deming, M. S. (2002). Design effects and intraclass correlation coefficients from a health facility cluster survey in Benin. *International Journal for Quality in Health Care*, *14*(6), 521-523.

Ryan, T. P. (2013). *Sample size determination and power*. Hoboken: Wiley.

Satterthwaite, F. E. (1946). An approximate distribution of estimates of variance components. *Biometrics Bulletin*, *2*(6), 110-114.

Schafer, J. L., & Graham, J. W. (2002). Missing data: Our view of the state of the art. *Psychological Methods*, *7*(2), 147-177.

Scheier, L. M., Griffin, K. W., Doyle, M. M., & Botvin, G. J. (2002). Estimates of intragroup dependence for drug use and skill measures in school-based drug abuse prevention trials: an empirical study of three independent samples. *Health Education & Behavior*, *29*(1), 85-103.

Schochet, P. Z. (2008). Statistical power for random assignment evaluations of education programs. *Journal of Educational and Behavioral Statistics*, *33*(1), 62-87.

Schouten, H. J. A. (1999). Sample size formula with a continuous outcome for unequal group sizes and unequal variances. *Statistics in Medicine*, *18*(1), 87-91.

Schulz, K., Altman, D., & Moher, D. (2010). Consort 2010 statement: updated guidelines for reporting parallel group randomised trials. *Journal of Clinical Epidemiology*, *63*, 834-840.

Schulz, K. F., Altman, D. G., Moher, D., & Fergusson, D. (2011). CONSORT 2010 changes and testing blindness in RCTs. *Lancet*, *375*, 1144-1146.

Scott, N. W., McPherson, G. C., Ramsay, C. R., & Campbell, M. K. (2002). The method of minimization for allocation to clinical trials: a review. *Controlled Clinical Trials*, *23*(6), 662-674.

Searle, S. R., Casella, G., & McCulloch, C. E. (1992). *Variance components*. New York: Wiley.

Senn, S. (1998). Some controversies in planning and analysing multi-center trials. *Statistics in Medicine*, *17*(15-16), 1753-1765.

Senn, S. (2002). *Cross-over trials in clinical research*. Chichester: Wiley.

Siddiqui, O., Hedeker, D., Flay, B. R., & Hu, F. B. (1996). Intraclass correlation estimates in a school-based smoking prevention study. *American Journal of Epidemiology*, *144*(4), 425-433.

Singer, J. D. (1998). Using SAS PROC MIXED to fit multilevel models, hierarchical models, and individual growth models. *Journal of Educational and Behavioral Statistics*, *23*(4), 323-355.

Singer, J. D., & Willett, J. B. (2003). *Applied longitudinal data analysis: Modeling change and event occurrence*. Oxford: Oxford University Press.

Slymen, D. J., Elder, J. P., Litrownik, A. J., Ayala, G. X., & Campbell, N. R. (2003). Some methodologic issues in analyzing data from a randomized adolescent tobacco and alcohol use prevention trial. *Journal of Clinical Epidemiology*, *56*(4), 332-340.

Slymen, D. J., & Hovell, M. F. (1997). Cluster versus individual randomization in adolescent tobacco and alcohol studies: illustrations for designs decisions. *International Journal of Epidemiology*, *26*(4), 765-771.

Smeeth, L., & Ng, E. S. W. (2002). Intraclass correlation coefficients for cluster randomized trials in primary care: data from the MRC trial of the assessment and management of older people in the community. *Controlled Clinical Trials*, *23*(4), 409-421.

Smith, P. J., & Morrow, R. H. (1991). *Methods for field trials of interventions against tropical diseases: A 'toolbox'.* Oxford: Oxford University Press.

Snijders, T. A. B. (2001). Sampling. In A. H. Leyland & H. Goldstein (Eds.), *Multilevel modelling of health statistics* (p. 159-174). New York: Wiley.

Snijders, T. A. B., & Bosker, R. J. (1994). Modeled variance in two-level models. *Sociological Methods & Research*, *22*(3), 342-363.

Snijders, T. A. B., & Bosker, R. J. (2012). *Multilevel analysis: An introduction to basic and advanced multilevel modeling.* London: Sage.

Spiegelhalter, D. J. (2001). Bayesian methods for cluster randomized trials with continuous responses. *Statistics in Medicine*, *20*(3), 435-452.

Spiegelhalter, D. J., Abrams, K. R., & Myles, J. P. (2004). *Bayesian approaches to clinical trials and health-care evaluation.* Chichester: Wiley.

Spybrook, J., Bloom, H., Congdon, R., Hill, C., Martinez, A., & Raudenbush, S. (2011). *Optimal design plus empirical evidence: Documentation for the optimal design software.*

Steel, R. G. D., & Torrie, J. H. (1980). *Principles and procedures of statistics: A biometrical approach* (2nd ed.). New York: McGraw-Hill.

Stein, A. C. (1945). A two-sample test for a linear hypothesis whose power is independent of the variance. *Annals of Mathematical Statistics*, *16*(3), 243-258.

Taljaard, M., Donner, A., Villar, J., Wojdyla, D., Velazco, A., Bataglia, V., ... Acosta, A. (2008). Intracluster correlation coefficients from the 2005 WHO global survey on maternal and perinatal health: implications for implementation research. *Paediatric and Perinatal Epidemiology*, *22*(2), 117–125.

Tan, F. E. S., & Berger, M. P. F. (1999). Optimal allocation of time points for the random effects model. *Communications in Statistics, Simulation and Computation*, *28*(2), 517-540.

Teerenstra, S., Eldridge, S., Graff, M., De Hoop, E., & Borm, G. (2012). A simple sample size formula for analysis of covariance in cluster randomized trials. *Statistics in Medicine*, *31*(20), 2169-2178.

Teerenstra, S., Melis, R. J. F., Peer, P. G. M., & Borm, G. F. (2006). Pseudocluster randomization dealt with selection bias and contamination in clinical trials. *Journal of Clinical Epidemiology*, *59*(4), 381-386.

Teerenstra, S., Moerbeek, M., Melis, R. J. F., & Borm, G. (2007). A comparison of methods to analyse continuous data from pseudo cluster randomized trials. *Statistics in Medicine, 26*(22), 4100-4115.

Teerenstra, S., Moerbeek, M., Van Achterberg, T., Pelzer, B. J., & Borm, G. F. (2008). Sample size calculations for 3-level cluster randomized trials. *Clinical Trials, 5*(5), 486-495.

Thomas, L. (1997). Retrospective power analysis. *Conservation Biology, 11*, 276-280.

Thompson, D. M., Fernald, D. H., & Mold, J. W. (2012). Intraclass correlation coefficients typical of cluster-randomized studies: estimates from the Robert Wood Johnson Prescription for Health Projects. *Annals of Family Medicine, 10*, 235-240.

Thompson, W. A. (1962). The problem of negative estimates of variance components. *Annals Mathematical Statistics, 33*(1), 273-289.

Torgerson, D. J. (2001). Contamination in trials: is cluster randomisation the answer? *British Medical Journal, 322*(7278), 355-357.

Torgerson, D. J., & Torgerson, C. J. (2008). *Designing randomised trials in health, education and the social sciences: An introduction.* Basingstoke: Palgrave Macmillan.

Tranmer, M., & Steel, D. G. (2001). Ignoring a level in a multilevel model: evidence from UK census data. *Environment and Planning A, 33*(5), 941-948.

Tsang, R., Colley, L., & Lynd, L. D. (2009). Inadequate statistical power to detect clinically significant differences in adverse event rates in randomized controlled trials. *Journal of Clinical Epidemiology, 62*(6), 609-616.

Turner, R. M., Omar, R. Z., & Thompson, S. G. (2001). Bayesian methods of analysis for cluster randomized trials with binary outcome data. *Statistics in Medicine, 20*(3), 453-472.

Turner, R. M., Prevost, A. T., & Thompson, S. G. (2004). Allowing for imprecision of the intracluster correlation coefficient in the design of cluster randomized trials. *Statistics in Medicine, 23*(8), 1195-1214.

Turner, R. M., Thompson, S. G., & Spiegelhalter, D. J. (2005). Prior distributions for the intracluster correlation coefficient, based on multiple previous estimates, and their application in cluster randomized trials. *Clinical Trials, 2*(2), 108-118.

Van Breukelen, G. J. P. (2013). ANCOVA versus change from baseline in nonrandomized studies: the difference. *Multivariate Behavioral Research, 48*, 895-922.

Van Breukelen, G. J. P., & Candel, M. J. J. M. (in press). Efficient design of cluster randomized and multicentre trials with unknown intraclass correlation. *Statistical Methods in Medical Research.*

Van Breukelen, G. J. P., Candel, M. J. J. M., & Berger, M. P. F. (2007). Relative efficiency of unequal versus equal cluster sizes in cluster randomized and multicentre trials. *Statistics in Medicine, 26*(13), 2589-2603.

Van Den Heuvel, E. T. P., De Witte, L. P., Nooyen-Haazen, I., Sanderman, R., & Meyboom-De Jong, B. (2000). Short-term effects of a group support program and an individual support program for caregivers of stroke patients. *Patient Education and Counseling*, *40*(2), 109-120.

Van Den Noortgate, W., Opdenakker, M.-C., & Onghena, P. (2005). The effects of ignoring a level in multilevel analysis. *School Effectiveness and School Improvement*, *16*(3), 281-303.

Van Schie, S., & Moerbeek, M. (2014). Re-estimating sample size in cluster randomised trials with active recruitment within clusters. *Statistics in Medicine*, *33*(19), 3253-3268.

Varnell, S. P., Murray, D. M., Janega, J. B., & Blitstein, J. L. (2004). Design and analysis of group-randomized trials: A review of recent practices. *American Journal of Public Health*, *94*(3), 393-399.

Veerus, P., Fischer, K., Hakama, M., & Hemminki, E. (2012). Results from a blind and a non-blind randomised trial run in parallel: experience from the Estonian postmenopausal hormone therapy (EPHT) trial. *BMC Medical Research Methodology*, *12*, 44.

Vermunt, J. K., & Magidson, J. (2005). Classification: The ubiquitous challenge. In C. Weihs & W. Gaul (Eds.), (p. 240-247). Heidelberg: Springer.

Vickers, A. J. (2003). Underpowering in randomized trials reporting a sample size calculation. *Journal of Clinical Epidemiology*, *56*(8), 717-720.

Vierron, E., & Giraudeau, B. (2007). Sample size calculation for multicenter randomized trial: taking the center effect into account. *Contemporary Clinical Trials*, *28*(4), 451-458.

Vierron, E., & Giraudeau, B. (2009). Design effect in multicenter studies: gain or loss of power? *BMC Medical Research Methodology*, *9*, 39.

Visser-Van Balen, H., Geenen, R., Moerbeek, M., Stroop, R., Kamp, G., Huisman, J., ... Sinnema, G. (2005). Psychosocial functioning of adolescents with idiopathic short stature or persistent short stature born small for gestational age during three years of combined growth hormone and gonadotropin-releasing hormone agonist treatment. *Hormone Research*, *64*(264), 77-87.

Walleser, S., Hill, S. R., & Bero, L. A. (2011). Characteristics and quality of reporting of cluster randomized trials in children: reporting needs improvement. *Journal of Clinical Epidemiology*, *64*(12), 1331-1340.

West, B. T., & Galecki, A. T. (2011). An overview of current software procedures for fitting linear mixed models. *The American Statistician*, *65*(4), 274-282.

Westine, C. D., Spybrook, J., & Taylor, J. A. (2014). An empirical investigation of variance design parameters for planning cluster- randomized trials of science achievement. *Evaluation Review*, *37*(6), 490-519.

Widenhorn-Muller, K., Hille, K., Klenk, J., & Weiland, U. (2008). Influence of having breakfast on cognitive performance and mood in 13- to 20-year-old high school students: results of a crossover trials. *Pediatrics*, *122*(2), 279-284.

Williamson, M. K., Pirkis, J., Pfaff, J. J., Tyson, O., Sim, M., Kerse, N., ... Almeida, O. P. (2007). Recruiting and retaining GPs and patients in intervention studies: the DEPS-GP project as a case study. *BMC*

Medical Research Methodology, *7*, 42.
Wittes, J., & Brittain, E. (1990). The role of internal pilot studies in increasing the efficiency of clinical trials. *Statistics in Medicine*, *9*(1), 65-72.
Wittes, J., Schabenberger, O., Zucker, D., Brittain, E., & Proschan, M. (1999). Internal pilot studies I: type I error rate of the naive t-test. *Statistics in Medicine*, *18*(24), 3481-3491.
Woertman, W., De Hoop, E., Moerbeek, M., Zuidema, S. U., Gerritsen, D. L., & Teerenstra, S. (2013). Stepped wedge designs could reduce the required sample size in cluster randomized trials. *Journal of Clinical Epidemiology*, *66*(7), 752-8.
Xu, Z., & Nichols, A. (2010). *New estimates of design parameters for clustered randomization studies findings from North Carolina and Florida* (Technical Report). National Center for Analysis of Longitudinal Data in Educational Research.
Yelland, L. N., Salter, A. B., Ryan, P., & Laurence, C. O. (2011). Adjusted intraclass correlation coefficients for binary data: methods and estimates from a cluster-randomized trial in primary care. *Clinical Trials*, *8*(1), 48-58.
Zhu, P., Jacob, R., Bloom, H., & Xu, Z. (2012). Designing and analyzing studies that randomize schools to estimate intervention effects on student academic outcomes without classroom-level information. *Educational Evaluation and Policy Analysis*, *34*(1), 45-68.

Author Index

Subject Index

Printed in the United States
by Baker & Taylor Publisher Services